The British Army and the Theory of Armored Warfare, 1918–1940

The British Army and the Theory of Armored Warfare, 1918–1940

Robert H. Larson

Newark: University of Delaware Press
London and Toronto: Associated University Presses

Associated University Presses
440 Forsgate Drive
Cranbury, N.J. 08512

Associated University Presses
25 Sicilian Avenue
London WC1A 2QH, England

Associated University Presses
2133 Royal Windsor Drive
Unit 1
Mississauga, Ontario L5J 1K5, Canada

Library of Congress Cataloging in Publication Data

Larson, Robert H., 1942–
The British Army and the theory of armored warfare, 1918–1940.

Bibliography: p.
Includes index.
1. Mechanization, Military—Great Britain—History.
2. Great Britain. Army—History—20th century.
3. Tank warfare—History. I. Title.
UG446.5.L293 1983 358′.18′0941 83-47509
ISBN 0-87413-219-3

Printed in the United States of America

To my children
Chris
Beth
Herb

Contents

Charts

Acknowledgments

The author of any extended work of scholarship is the beneficiary of innumerable acts of assistance and kindness which, if taken in their entirety, are essential if the project is ever to be completed. I am no exception. Since I first began to investigate the evolution of British military doctrine for a modest seminar paper in modern British history at the University of Virginia, I have received both material assistance and moral support from numerous people on both sides of the Atlantic Ocean. It is with a great deal of pleasure that I now am able to publically acknowledge this debt.

At the University of Virginia, I wish to thank Dr. Nicholas C. Edsall, my major advisor, and Dr. Martin J. Havran, my second reader, for their thoughtful guidance and careful criticism as this work was going through the dissertation stage and, after its acceptance, for their encouragement to prepare it for publication.

At Lycoming College, I wish to thank the administration for two grants to assist covering the costs of research and typing. I also wish to acknowledge my great debt to my colleagues in the History Department who have worked to create an environment where teaching and research go hand in hand and where academic collegiality is truly a way of life. In particular, Drs. Loring B. Priest and Robert H. Ewing, both now Professors Emeriti, read and made numerous valuable suggestions on portions of early drafts of this work, and Drs. John F. Piper and Richard J. Morris listened to my problems and offered many helpful suggestions. Dr. Morris also provided invaluable support as a professional critic and personal friend during the final, hectic days before this manuscript was submitted to the publisher. His great store of energy and unfailing sense of humor went far to preserve a degree of sanity at a time when, by all rights, there should have been none.

In England, I wish to express my deep gratitude to Lady Kathleen Liddell Hart for her permission to have full and free access to her husband's papers, then stored at her home in Buckinghamshire, and for her kind hospitality while I was working there. It is scarcely an overstatement to say that without her assistance this book would not have been possible or that without her kindness a considerable portion of the research would not have been nearly so pleasant. I also wish to thank the staffs of the Public Records

Office and the Sir Basil Liddell Hart Center for Military Archives at Kings College, London, for their assistance.

At the U.S. Army Military History Institute, Carlisle Barracks, Pa., I wish to thank Colonel George S. Pappas, U.S. Army, Ret., Mrs. Joyce Gooding (née Eakin), Drs. Benjamin F. Cooling and Richard J. Somers and Mr. John Slonaker and to acknowledge the assistance of the late Colonel James B. Agnew. Their help in so many ways, both large and small, has saved me countless hours of fruitless searching.

I also wish to thank Miss Trudy Phillips and Mrs. Esther Henninger for typing various drafts of the manuscript and Mrs. Sandy Rife for preparing the organizational charts.

Finally, I want to give a very special thanks to my wife, Mary Ellen, whose encouragement and support during the final stages of work on this book were matched only by her skills as research librarian, writing critic, and editor. Indeed, without her understanding, attention to detail, and insistence on meeting deadlines, this work might not yet be seeing the light of day.

The British Army and the Theory of Armored Warfare, 1918–1940

• 1 •

Years of Defeat: The Search for Responsibility

THE British Army's inability to counter the German blitzkrieg during the opening campaigns of World War II in France and North Africa stands out as one of the great ironies of that conflict. The British, after all, had invented the tank. They had pioneered its development during World War I, employing it with awesome effectiveness in the latter stages of that conflict, and had produced the leading theorists of armored warfare during the interwar years in J. F. C. Fuller and B. H. Liddell Hart. Yet by 1939 this early lead had been lost, and for the ensuing three years the British Army paid a heavy toll in lives and equipment before it was able to reverse the tide.

Until now, historians have ascribed this turn of events to social conservatism within the officer corps and shortsighted government policy, either or both of which, according to whom one believes, dominated the army during the interwar years and produced its state of unpreparedness in 1939. The army, according to those advocating the first explanation, was dominated by officers who continued to retain an archaic faith in the horse as an instrument of war despite the lessons of World War I. With a stubborness worthy of Wellington's lines at Waterloo, these officers blocked both the expansion of the Royal Tank Corps and the mechanization of the cavalry. Other historians, however, have emphasized that successive British governments were responsible for the army's stagnation by severely restricting military spending and by adopting a strategic policy that limited its mission to policing the empire where tanks would be of little value. Finally, there are those who have concluded that both factors played a role: the army had no desire to develop either a powerful tank arm or a coherent doctrine of tank warfare, and the government withheld both the means and the rationale to do so. Unfortunately, despite their widespread acceptance, the apparent logic and evidence supporting these explanations are much less impressive when closely scrutinized.

Horses versus Tanks

Few caricatures of the British Army have achieved the popularity of Robert Low's Colonel Blimp, that plump, pompous proponent of stagnation who steadfastly opposed the introduction of fresh ideas throughout the 1930s. The cartoon itself has long-since disappeared from British periodicals, but the belief persists that such officers typified the army at that time and that their influence explains its failure to meet the challenge of the blitzkrieg. Essentially this stereotype rests on three assumptions: first, that the officer corps was dominated by the "landed classes," who entered the army seeking a leisurely outdoor style of life compatible with their upbringing; second, that because of this background, they held the cavalry in special esteem, enabling it to secure a position from which it could significantly influence the army's future; and, third, that the cavalry, aided and abetted by the army as a whole, staunchly opposed the expansion of the armored forces as a threat to their traditional way of life. Both as a participant and as a historian of these events, Sir Basil Liddell Hart often stressed the cavalry's opposition to his ideas, and in his *Memoirs* wrote:

> Cavalry circles loathed the idea of giving up the horse, and thus instinctively decried the tank. They found much support both in the War Office and in Parliament. Wellington's reputed saying that the Battle of Waterloo was won on the playing-fields of Eton is merely a legend, but it is painfully true that the early battles of World War II were lost in the Cavalry Club.[1]

With some modifications and varying degrees of emphasis, other historians have followed this line of argument closely. In his biography of J. F. C. Fuller, for example, Anthony J. Trythall ascribed many of his subject's difficulties to the "lack of any widespread acceptance in many regiments that the days of the horse and of marching infantry were over and numbered respectively."[2] Likewise, Kenneth Macksey, in his biography of "Patrick" Hobart, the outspoken commander of the First Tank Brigade from 1934 to 1937, emphasized the difficulties of convincing the army of the need for change when many officers "looked on their military careers as a pleasant occupation, about which, to say the least, they could become detached."[3] Correlli Barnett, on the other hand, perceived a more subtle consequence of the social background of the officer corps. In his account of the North African campaign, *The Desert Generals,* Barnett agreed that most British officers opposed mechanization because their rural upbringing gave them little understanding of modern, industrial society or war, but he further claimed that this provincialism also led many—specifically in the infantry and the cavalry—to focus their loyalties and professional outlook on their individual regiments. When the army finally accepted the replacement of horses by tanks, it sought to protect this regimental structure by mechanizing the cavalry rather than expanding the Royal Tank Corps, justifying this decision with the "fallacious analogy" that this was merely a change in the mode of transportation and not a change in the units' mission. They were abetted in this by many Royal Tank Corps officers who, after years of discrimination and misuse, fell prey to extremists within their own ranks who argued that armor would only achieve its full potential if allowed to operate completely on its own. As a result, the combined arms concept of close cooperation among infantry, armor, and artillery was ignored and the defeats in North Africa followed inevitably.[4] In short, Barnett believes that the social composition of the officer corps not only delayed the expansion of Great Britain's tank forces—in a later book he specifically stated it took place only after 1937 and against the "bitter rearguard opposition of the traditionalists"[5]—but also prevented the establishment of a sound tactical and strategic doctrine for their employment in combat.

Any assessment of this "social" interpretation must begin with the observation that the "landed classes" had long since lost their numerical domination of the officer corps by the interwar years. From about the middle of the nineteenth century, they had been replaced by the middle class at an accellerating rate, so that by 1930 the latter held an absolute majority both in the officer corps as a whole and in its senior ranks. In 1830, for example, members of the aristocracy and landed gentry constituted fifty-three percent of the entire officer corps and eighty-nine percent of those holding the rank of major general or above; by 1912, they held only forty-one percent of all commissions and sixty-four percent of the positions of major general or above, and by 1930 their proportion had declined to eleven percent and forty percent of those categories respectively.[6] Indeed, one authority, noting that both the prewar military radicals and the successful British commanders of World War II came from the same social background as the army hierarchy of the interwar period, has concluded that the "social interpretation" is "very largely nonsense."[7]

The second assumption, that the cavalry was held in such esteem within the army that it had the power to affect significantly the army's future, is much more difficult to evaluate precisely. The prestige of the combat arms is often associated with popular folk myths and stereotypes exemplified by the British infantry's image as the indomitable thin red line or the cavalry's bravado reflected in Tennyson's "The Charge of the Light Brigade." Certainly these images contribute to this prestige, but they cannot be measured nor can they be used to demonstrate the extent to which the army itself placed value on a given combat arm. A more valuable approach can be found in the promotion patterns of officers from the ranks of regimental duty, where service is confined to a single arm, to the general-officer ranks, where service is not normally identified with a particular arm but, rather, with commanding units composed of all arms. A study of these patterns would reveal the extent to which the army perceived each arm as suitable training for such positions of high command, which, in turn, would provide a strong indication of the relative value placed on each arm.

For the British Army of the interwar period, such a study reveals a significant shift of values as indicated by the following figures[8]:

		Infantry		*Cavalry*		*Artillery*	
Colonels	1923	56%	(190)	6%	(20)	38%	(129)
	1932	56.3%	(146)	6.6%	(17)	37%	(96)
	1938	51%	(134)	8.1%	(21)	40.4%	(105)
All Generals	1923	70.8%	(73)	9.7%	(10)	19.4%	(20)
	1932	62.8%	(54)	7%	(6)	30%	(26)
	1938	61.3%	(46)	4%	(3)	34.7%	(26)
Lieutenant	1923	75%	(21)	17.8%	(5)	7.1%	(20)
Generals	1932	59.2%	(16)	7.4%	(2)	33.3%	(9)
and above	1938	56.5%	(13)	0%	(0)	43.5%	(10)

It is clear from these figures that, while the proportion of colonels in the three combat arms did not change significantly relative to each other during the interwar years (which is to say, the size of pool from which the generals were chosen remained the same), the proportion of cavalry colonels promoted into the general-officer ranks steadily declined, while the proportion of artillery colonels promoted steadily increased. These trends are accentuated even more sharply when the major generals are excluded from the figures: between 1923 and 1938, cavalrymen were gradually eliminated from this category while the number of artillerymen increased sixfold. What is clear from these figures is that the infantry and artillery were moving steadily toward a position of equality in the senior ranks of the army while the cavalry was being steadily relegated to a position of insignificance, if not total exclusion. This decline is even more striking when compared to the rise of Royal Tank Corps colonels to the rank of major general.[9]

		Cavalry	*Royal Tank Corps*
Colonels	1923	20	0
	1932	17	11
	1938	21	15
Major Generals	1923	5	0
	1932	4	0
	1938	3	4

The almost total absence of cavalrymen from the inner circles of the army's power structure is equally striking. Throughout the interwar years, the British Army was ruled by a body known as the Army Council. A product of the post–Boer War reforms, it was chaired by the secretary of state for war, a member of the cabinet, and was composed of the three civilian heads of departments and four military members: the chief of the imperial general staff (CIGS), who was responsible for training and preparation for war; the adjutant general (AG), who was responsible for recruiting, medical, and personnel services; the quartermaster general (QMG), who was responsible for housing and all supplies except weapons and ammunition; and the master general of ordnance (MGO), who was responsible for the design and production of the last two items.[10] During this period, a total of twenty-four appointments was made to these positions and to the position of director general of the territorial army. Of them, fourteen went to infantrymen, seven to artillerymen, two to engineers, and only one to a cavalryman. Moreover, the single cavalryman, General (later Field Marshal) Sir Phillip W. Chetwoode, served for only six months (August 1922 to February 1923) before moving to another post.[11]

The conclusions to be drawn from this evidence are clear: first, during the interwar years the British Army increasingly came to believe that service with the cavalry was less important than service with the other combat arms when it came to training officers for senior positions of responsibility; and, second, at no time did the cavalry hold a position of power from which it could control or even significantly influence the army's future. The explanation for this conclusion, although it cannot be documented fully, also seems clear. Certainly it was not the result of any directive or consciously adopted policy by the army. It is logical, however, that, given the experience of World War I, cavalry officers' wartime service would not be so impressive as that of officers from the infantry or artillery and they would, therefore, be less likely to be promoted. In addition, as will be recounted in more detail later, the importance of the cavalry's role in military operations was steadily reduced in the *Field Service Regulations* during the interwar year, lending further substance to its declining influence. Finally, there is evidence that, in some quarters at least, cavalry officers were viewed as less capable professionally than their counterparts in the other combat arms. Although it is impossible to determine the extent of this attitude, the evaluation of a cavalry training exercise in 1935 by General Sir John Burnett-Stuart, an infantryman then holding one of the major commands in the British Isles, was most likely not unique:

> This exercise confirmed my previous conviction that the Cavalry are still behind the other arms in their conception of war, in professional leadership and in the thoroughness of their training. They are inclined to forget that the Cavalry spirit requires very professional technique as its medium of expression.[12]

The third assumption, that the cavalry staunchly opposed mechanization as a threat to its traditional way of life, is the most difficult of the three to evaluate. In the first place, it rests on the belief that the army officer corps consisted overwhelmingly of officers who had little sense of professionalism. Certainly this was true in the late nineteenth century, when most senior officers of the interwar period began their careers. One officer, recalling these years, later wrote that officers were not called upon to work long hours, but sufficient only to add zest to other pursuits and to make life pleasant.[13] Accounts by his contemporaries reveal that it did not take a great deal of work to add the desired zest. The duty day was normally over by eleven in the morning unless the unit was going hunting, in which case it was often over by breakfast. In some units officers could expect to go hunting as often as six days a week and even to hire government horses for the purpose. Incentives for self-study, moreover, were almost totally lacking and sometimes such activity even was discouraged by senior officers.[14] No wonder that a Royal Commission studying the army in 1902 concluded that "Keenness is out of fashion; it is not correct form: the spirit and fashion is rather not to show keenness."[15] No wonder, too, that the British Army performed so poorly in the Boer War or that its own official history of that conflict admitted, "As a fighting machine it was a sham."[16]

The glaring deficiencies revealed by that war and the growing possibility of a confrontation with Germany in the decade before 1914, however, wrought a dramatic change in this state of affairs. A wave of professionalism swept the army that, according to one authority, had not existed since the days of Cromwell, and the Staff College "became the only sure means of advancement."[17] The college itself was in the forefront of the new spirit. Beginning with the appointment of Henry Rawlinson as commandant in 1903, the curriculum was revitalized and modernized. Rawlinson was succeeded in turn by Henry Wilson and William Robertson, both of whom maintained the high standards he had established. By 1913, the prestige of the Staff College was such that 185 officers applied for the thirty-six vacancies open for competition.[18] Yet, the old ways by no means died out. Despite the new prestige of the Staff College and its graduates, there is undoubtedly much truth in the conclusion of one noted British military historian that "a large number of officers—perhaps even a majority—remained unconvinced [of the need for military professionalism] right up to 1914."[19]

World War I shattered this comfortable illusion, but it also revealed that the army had produced a core of highly proficient officers capable of transforming a professional army of a few hundred thousand into a national army of millions. In 1914, the British Regular Army at home consisted of only one cavalry and six infantry divisions and lacked provisions for raising troops on anything approaching the Continental scale. The largest force it had ever fielded during the lifetime of any officers then serving had been the relatively small number raised for the Boer War.[20] Despite this absence of machinery and experience, however, the British Expeditionary Force in France alone numbered fifty-two British and nine dominion divisions by August 1916.[21] This ten-fold expansion, which took place in a time of war and with remarkably little confusion, cannot be dismissed as an insignificant accomplishment. The administrative and logistical expertise required to train, organize, equip, dispatch, and supply a force of this magnitude by officers without prior experience is a clear demonstration that the efforts of the Staff College had not been completely in vain.

Unfortunately a reaction against the strains and demands of the war set in almost immediately after 1918. Correlli Barnett, in fact, has argued that the horizon of the officer corps quickly reverted not to 1914, but to the 1890s, when professionalism had been equated with "vulgar careerism."[22] Once again promotions were based almost solely on seniority, not only eliminating one of the prime incentives for professional development, but also producing blatant inequities. By 1924, the problem had become so severe that the secretary of state for war publicly announced that the principle of "promotion by merit" would strictly be followed in the future.[23] Even the Staff College was unable to disassociate itself completely from this trend. Writing a history of that institution in 1927, one officer even went so far as to make this into a virtue. "If there is any specially sterling characteristic of life at the Staff College," he wrote, "one would say it is the peculiarly British spirit of humor bubbling through it. Everyone is expected to do a job of work without making heavy weather, or taking it, or himself, too seriously." He further noted that "sport remains as marked a characteristic of the life there as ever. . . . The Drag [hunt] remains the mainstay of the life of the Staff College, and it will be a sad day if a mechanical age ever renders its continuance difficult."[24] This situation eventually caused enough concern that a special War Office committee, examining the need for the further training of senior officers beyond the Staff College, hopefully concluded that such an institution would "go far to banish the ingrained British idea that continued presence with troops, chance participation in Small Wars, and exercise in games are the best training for high command."[25]

To end a description of the British officer corps at this point, however, would be to ignore a good deal of contradictory material indicating that professionalism was by no means dead in the army at this time. The very fact that there were officers who believed that there was a need for education beyond the Staff College, for example, demonstrates that not all officers believed that service with troops, experience with occasional colonial expeditions, and "exercise in games" were the best prerequisites for high command. Throughout the interwar years, moreover, the British Army supported five major professional journals devoted to the study of military history and affairs. These journals—*Journal of the Royal United Service Institution* (which was interservice), *Army Quarterly, Cavalry Journal, Journal of the Royal Artillery,* and *Journal of the Royal Engineers*—were far more than regimental publications devoted to promotions, sports, and social life. All were seriously interested in the lessons to be learned from the recent war and in the strategic problems facing the British Empire. They were also concerned with the future course of war and particularly with the impact mechanization would have on the army. On a number of occasions, they sponsored essay competitions inviting officers to contribute their views on this subject. Naturally, not all of the articles published during these years were of high quality, and not all views expressed were progressive. The great majority, however, reflected a good deal of thought on the part of the authors and the leading tank advocates were able to make considerable use of the journals as a forum for their views. In short, while it is probably true that there continued to be an element within the officer corps that exhibited a disturbing detachment from its profession, it would be patently wrong to claim that professionalism was dead or, more importantly, that the only officers who were interested in professional development existed in the camp of the tank advocates.

The heart of this social interpretation, however, rests on the belief that the army was blindly devoted to the retention of horsed cavalry and hostile to tanks. There were, to be sure, many officers devoted to the proposition that horsed cavalry could still play a vital role in modern war. As their most prominent spokesman, Field Marshal Haig often singled out the cavalry for praise in his dispatches covering operations in France and in the last concluded confidently:

> In the light of the full experience of the war the decision to preserve the Cavalry Corps has been completely justified. It has been proved that cavalry, whether used for shock effect under suitable conditions or as mobile infantry, have still an indispensable part to play in modern war. Moreover, it cannot safely be assumed that in all future wars the flanks of the opposing forces will rest on neutral states or impassible obstacles. Whenever such a condition does not . . . [exist] opportunities for the use of cavalry must arise frequently.[26]

Several years later, he condensed this opinion into a single sentence that is undoubtedly the most widely quoted statement of the entire mechanization controversy. "As time goes on," he wrote in 1925, "you will find just as much use for the horse—the well bred horse—as you have ever done in the past."[27] Others repeated the same message. Speaking in Parliament in 1921, a member of Parliament and former officer maintained that "I believe the right honorable gentleman [the secretary of state for war] and his advisors are entirely wrong in their thinking that they can substitute tanks for cavalry. Every advance in science has made the horse a more and more indispensable weapon of war."[28] Even more outlandish were the comments by an unknown individual who wrote in the April 1934 issue of the *Army Quarterly:*

> Personally, I feel that the chief reason why certain people dislike getting on to a horse is that they are afraid; they consider the brute a nasty, unsafe form of convenience. They may be quite calm and brave in many

> other situations but there is a kink in the armour: we do not want kinks in the armour of our leaders; if the leaders are "nervy" about a horse, they are usually nervy about other things. . . .
>
> Looking through the list of those who were in my batch at the Staff College just before the war, I note the names of several officers who were perniciously bad riders: not one of them made good on "active service," whereas nearly all those who rode well became leaders and several of them now hold high ranks in the Army. As I write, I think of all the generals I have served under or who I know about. I cannot think of one who has been a success in war who has not been a good horseman; nor can I think of the name of any general who "hated the sight of a horse," who has been a real leader of men. There may be such but I have not heard of them.[29]

Most arguments for the retention of horsed cavalry were far more reasoned than this one, however. Cavalrymen took particular comfort in Haig's assessment of their role in the closing days of the war in France and claimed that it was only their action in pursuing the retreating Germans and continuously driving in their rearguards that prevented the enemy from reestablishing a strong defensive position.[30] They were even more hypnotized by Field Marshal Allenby's 1918 Palestinian campaign that virtually annihilated a Turkish army. In that campaign, Allenby had used his cavalry first to penetrate deeply behind the main Turkish positions, cutting off much of their forces from their lines of retreat, and then to pursue the remnants that managed to slip through. To many cavalrymen this was proof that they could still play an essential role in modern war under "normal" conditions. They claimed that they alone possessed the mobility, maneuverability, and endurance to carry out such an operation. Infantry could not hope to match their speed, and although tanks were not present in Palestine, the cavalry felt sure that they could not have done the job either. Tanks, they argued, lacked the necessary speed, would have had considerable difficulty covering some of the more treacherous terrain, and would have created insurmountable logistical problems. Although the tank advocates responded that much of the cavalry's success was the result of severe arms and equipment shortages among the Turkish forces, there was no denying the cavalry this moment of glory and it gripped all those who continued to see that arm in its traditional role.[31]

Cavalry advocates, however, did not restrict their vision to the past. One officer, who had served with both horsed and mechanized units, justified the continued existence of the former, citing their ability to perform reconnaissance, flanking, screening, and pursuit operations in all types of climate and terrain and without a complex supply system. He admitted that under certain conditions tanks might perform their missions equally well, but the crucial advantage of the cavalry was its ability to accomplish these tasks under any conditions. This advantage, he continued, would increase in importance in the future as air power put a premium on movement during periods of poor visibility. At some future date, he concluded, these problems might well be solved but, rather than oppose such changes, he declared that "no one will be better pleased than the cavalryman to see the horse spared the miseries of the battlefield. Unfortunately, that day still seems far distant."[32]

As these last remarks indicate, many of those who believed that cavalry still had a role in future war were also skeptical, if not hostile, to tanks. Such men asserted that tanks were a product of the Western Front, which they held to be an aberration in the evolution of war, and they argued that it was dangerous for the army to plan its future on such a narrow experience when it might well have to fight under very different conditions. One such critic, for example, boldly labeled the tank a "freak," explaining that "the circumstances which called it into existence were exceptional and not likely to recur. If they do," he concluded, "they can be dealt with by other means."[33] Another, who admitted that tanks had played a valuable role during the war, nonetheless claimed that their successes were primarily the result of their novelty, the absence of antitank guns, and other favorable conditions that would probably not exist in future wars. In the specific instance of the British assault on Amiens on August 8, 1918, he maintained that the terrain and weather were "so abnormally favorable to the tank arm that it might be a matter of doubt whether they will recur again."[34]

Such critics were also singularly unimpressed by arguments that advances in technology would enable tanks to overcome their teething difficulties. They specifically called attention to the problems of tank design, the power of antitank weapons, the restrictions of terrain on tank mobility, and the difficulties of supplying large numbers of tanks with essentials such as fuel and ammunition. In each of these cases they argued that advances in one of these areas (such as new tank designs that would increase mobility) would only add to problems in other areas (such as maintaining supply lines). As one critic concluded:

> Given favorable conditions, the tank can play a useful if expensive role in exploiting the superiority of the other arms and leadership in general. It cannot compensate for the lack of such superiority, nor can it act independently of the support of the other arms.[35]

Yet, as with the question of the state of professionalism in the officer corps, there is considerable evidence to demonstrate that this portrait of army attitudes is also serious oversimplification. The British Army was certainly neither so enthusiastic about tanks as Fuller and Liddell Hart nor so ready as they were to reduce or abolish the cavalry. But to allow the foregoing statements to stand as the army's final word on the subject would be both unfair and innaccurate.

Field Marshal Haig, for example, was no less generous in praising the Tank Corps for its contribution to the final Allied victory than he had been in praising the cavalry. In his dispatch covering the final months of the war, he wrote:

> Since the opening of our offensive on August 8, tanks have been employed on every battlefield and the importance of the part played by them in breaking up the resistance of the German infantry can scarcely be exaggerated. The whole scheme of the attack of August 8 was dependent upon tanks, and ever since that date on numberless occasions, the success of our infantry has been powerfully assisted or confirmed by their timely arrival. So great has been the effect produced upon the German infantry by the appearance of British tanks that in more than one instance, when for various reasons real tanks were not available in sufficient numbers, valuable results have been obtained by the use of dummy tanks painted on frames of wood and canvas.
>
> It is no disparagment of the courage of our infantry, or of the skill and devotion of our artillery, to say that the achievements of those essential arms would have fallen short of the full measure of success achieved by our armies had it not been for the very gallant and devoted work of the Tank Corps, under the command of Major General H. J. Ellis.[36]

Given Haig's image as a die-hard cavalryman and his comments on that arm noted earlier, this appreciation is striking, to say the least. Yet there is no inconsistency if each statement is taken on its own merit, that is to say, if they are taken to indicate that Haig believed both cavalry and tanks played a vital role in the final Allied victory. Such an explanation fits in well with the overall tenor of his dispatches. There can be no doubt that Haig never lost his faith in the value of cavalry, but it is equally true he never believed it or any other single arm could achieve military victory alone; that could only come through the cooperation of all arms.

Likewise, many of the tank's critics had no desire to abolish tanks from the British Army. One of those just noted, in fact, explicitly stated that his criticisms were directed at what he believed were the more extreme claims of the tank advocates and the hazards of predicting the future of war under peacetime conditions. Having made these criticisms, he nonetheless asserted that tanks were one of the four elements necessary for any successful attack. Together with artillery, machine guns, and riflemen, he called them an essential part of any offensive team and, like Haig, held success could only result from the cooperation of all arms.[37]

Although this public debate is a useful index of army opinion, its importance is far outweighed by the reports of official War Office committees that were formed to examine various problems facing the army. Composed of senior officers who held key positions either in the War Office or in the major commands in the British Isles, these committees normally were established by the Army Council on an ad hoc basis to examine a specific issue. As such, they are probably the best source available to determine the attitudes of the army establishment. Significantly, the reports of these committees reveal a growing awareness of the limitations of the cavalry and a growing acceptance that the days of the horse as an instrument of war were numbered.

The first such committee to delve into this area was established in 1926 under General Sir Archibald Montgomery-Massingberd to examine the army's cavalry requirements and determine if and where expenditures for this arm could be reduced. Although it was not apparently limited to any particular aspect of the cavalry's future, the committee felt itself restrained by the Ten-Year Rule, which restricted the army to policing the empire, and "the unce: tainty as to the lines along which tanks will develop."[38] Nonetheless, it produced a report that, although admittedly cautious, reveals a far more progressive attitude than the army is generally given credit for. The committee's first conclusion was clear and direct:

> It is the practically unanimous opinion of those officers who have been consulted by the Committee that the cavalry, as at present organized and equipped, has neither sufficient mobility nor sufficient fire-power to enable it to carry out its duties in modern war.[39]

It asserted, however, that this state of affairs was not the result of the obsoleteness of the horse, but to the excess weight of cavalry equipment, especially the machine guns, and the use of horse-drawn vehicles for first-line transport. Consequently, the committee recommended that both of these be mechanized as soon as possible and expressed confidence that these steps would enable the cavalry to perform its missions successfully.[40] The reaction to these two recommendations indicates that they had widespread support throughout the army, but this is of limited significance as neither change would affect the tactical role or importance of the horseman. Indeed, they were designed to support rather than replace him by removing the heavy incumberances that were restricting his mobility.

Of much greater importance was the committee's attitude toward the mechanization of the cavalry as a whole. Because of the uncertainties surrounding future tank design, the committee felt it could not examine the possibility of tanks' replacing horses, but it did believe alternatives existed in the form of armored cars and small, one or two-man tanks. In its discussion of these alternatives, the committee agreed that such vehicles enjoyed significant advantages over horses with regard to speed, radius of action, and small-arms fire, but it also found serious disadvantages in them as well. In particular, these vehicles had problems maneuvering over certain types of terrain, required highly trained crews, and needed a complex maintenance system. Far more important for the immediate future of the cavalry, however, was the fact that no vehicle then existed capable of meeting the army's requirements. The committee believed this was a serious deficiency, the correction of which deserved the highest priority, and recommended "that the design and production of a cross-country armored car . . . be regarded as a matter of extreme urgency." It further recommended that as soon as this vehicle was developed, probably no less than two years hence, it should be tested by the cavalry to determine its role, its effect on organization and equipment, and its ability to replace horses as the means of transportation; and that in the interim armored cars should replace horses on a limited basis in selected cavalry regiments. Finally, and perhaps most significantly, it recommended that the funds for this program be obtained by reducing the strength of each of the cavalry regiments by ten percent.[41]

Clearly this report was far short of being revolutionary. It sidestepped the question of the complete mechanization of the cavalry, and its recommendation to give armored cars to the cavalry rather than expand the Royal Tank Corps was admittedly done in part to assuage ruffled feelings. The committee had considered both alternatives and, while it agreed that the Royal Tank Corps already possessed the technical knowledge and experience to make expansion relatively easy, concluded that the tradition and esprit de corps of the cavalry were vital to the army and should be maintained.[42] Yet, the conservatism inherent in this decision can be overstressed, for even two years later Liddell Hart admitted that cavalry officers were ideally suited to lead mechanized formations as they, more than anyone else in the army, had preserved the spirit of mobility.[43]

These limitations notwithstanding, however, it is equally undeniable that the committee viewed the mechanization of the cavalry in an objective and generally positive light. Its reluctance to urge the complete abolition of horsed cavalry is understandable given the absence of an adequate replacement vehicle. Its willingness to reduce the size of the cavalry to provide funds for mechanization is striking. The importance of this recommendation cannot be ignored, for it demonstrates that by the mid-1920s, there was a significant element in the War Office willing to reduce the size of the cavalry to support mechanization. Given the extreme budgetary limitations under which the army then operated, the absence of any threat of the international scene, and the lack of a vehicle to meet the army's specifications, this was no mean achievement.

The army's response to the report can be determined only to a limited degree. The reaction of the members of the Army Council was generally favorable.[44] Within the cavalry, there is some evidence that they were more than satisfied with the outcome of the report, and Montgomery-Massingberd received two letters from cavalry officers to this effect. Both expressed deep appreciation for the considerable sympathy he had showed to the cavalry and declared that the recommendations would have the full support their arm. One of them, from the commandant of the Equitation School, confidently expressed the feeling that "if the powers that be will only work on the lines suggested in the Report, I am certain in my own mind that we will be worth every penny that is spent on us, if we really get to moving warfare."[45]

Another source of the army's attitude toward mechanization may be found in the reaction of cavalry units actually faced with this prospect. Although most seemed to have accepted this change without comment, there are two instances in which the units' reactions have been preserved.

The first reveals an attitude wholly in tune with the classical stereotype. In the spring of 1937, a rumor was circulated that the Scots Greys were to be mechanized. In response, the commander of the regiment informed various public officials in Scotland and Scottish members of Parliament of this possibility in the hope that something could be done at least to delay the event. Although this rumor was totally unfounded,[46] a fact that was not publicly known, the Scots reacted immediately and flooded the War Office with letters. A few opposed mechanization in principle, but most simply maintained that the regiment should be among the very last to be mechanized out of respect for its proud tradition. To do otherwise, they maintained, would "offend" Scotsmen wherever they might reside.[47] The Army Council's reaction, while sympathetic to regiment's traditions, was decidedly unfriendly to this attempt at outside interference. They viewed the actions of the Grey's commander as a breach of discipline and eventually reprimanded him.[48] As for the unit itself, it remained unmechanized for the time being, not because of the outcry raised, but because the mechanization program had not yet reached them. The intense feeling this change could raise is significant, but the fact that this was the only instance in which the mechanization of a cavalry regiment caused any outcry at all is no less significant. Whatever their feelings may have been, the other cavalry regiments accepted mechanization, if not with glee, then at least without overt opposition.

Quite probably, in fact, some even shared the more progressive attitude, the Third King's Own Hussars. At the end of 1934, the army decided to conduct a series of exercises determine the most feasible organization for mechanized cavalry regiments. The Third Hussars actually volunteered to carry them out, but with the understanding that this would not prejudice the unit's chances of remaining horsed. In his report of the tests conducted in the summer of 1935, the unit commander, Lieutenant Colonel Grubbe, informed the War Office that they had staged a demonstration of their new equipment to past officers of the regiment and reported:

> The consensus of opinion is one of the greatest regret of the loss of the horses, but at the same time they are unanimous in wishing the regiment to move with the times, and they consider that the eventual mechanization of the cavalry is inevitable in some form or another.[49]

In addition, Colonel Grubbe spoke with officers from allied regiments in Canada, Australia, and New Zealand, and reported that their reactions were more varied. The officer from Australia felt that mechanization was definitely the course of the future, but the other officers were not so sure. They were from rural areas where horses were still widely used in civil life and were highly dubious of the advantages of mechanization. Another officer, however, who had served on the Trans-Jordan frontier and in the

Sudan expressed the opinion that mechanized cavalry could have outperformed horsed units in all cases and in all weather.[50]

Given this divergence of opinion, what accounts for the continued belief that the cavalry was steadfastly hostile to tanks and the prospect of mechanization? Stereotypes arising out of the Scots Greys' incident certainly form part of the answer, but a far more important reason lies in the army's conviction that the horse was an important aid in training officers. This was particularly striking, for if the cavalry's attitude toward tanks can be described as cautiously progressive, its attitude toward the retention of horses for training officers can only be regarded as blindly obstinate. Indeed, General Pile, when noting this contradiction to the author, suggested that the cavalry would have readily accepted mechanization had they been assured they could keep their horses for hunting and other personal uses.[51] The problem was partially a matter of cost, but more importantly, it was a matter of whether cavalry officers could convert to a "mechanized state of mind" if they retained horses for personal use. This point was raised by one of the tank advocates who told Liddell Hart that "as long as you continue to provide the Cavalry officer with two free chargers, although his unit is equipped with tanks, so long psychologically will be be wedded to the horse and indifferent to or averse from getting down to the machine."[52]

The army, however, refused to accept this view. In 1935, a special War Office committee composed of the principal staff officers to the military members of the Army Council and one civilian (the assistant secretary of state for war) issued a paper entitled "Report of the Committee on the Value of the Horse in the Training of Officers," which dealt with this very question. In it, the position of the military members of the committee was clear and unequivocal: the best interests of the army demanded the retention of horses for the use of officers in mechanized units. They based this conclusion on two premises: first, that it was essential in fostering certain military qualities absolutely necessary for an officer such as mental alertness, ability to "read" ground accurately, ability to reach quick decisions, and physical fitness; and, second, that it was an important element in attracting the "right people" to the army. The military qualities, they argued, required training over unfamiliar ground, an impossibility if officers rode only in tanks because the damage caused by the vehicles would require that training be restricted to very limited area with which they could soon become familiar.[53] The second point is extremely interesting and reflects the state of professionalism within the officer corps previously noted. If these arguments were valid, it would be logical to have tank officers ride, and this was, not surprisingly, recommended on a restricted level. As for other mechanized units, the report recommended that the number of horses be reduced from two to one per officer, but established no timetable for this.[54] It is noteworthy that the only objection to this recommendation came from

H. J. B. Clough, the sole civilian member of the committee. He felt so strongly about this that he included his own minority report, in which he argued that there was no direct military need for horses and that they should therefore be eliminated from all mechanized units as soon as possible, as "the longer horses remain, the more they will come to be regarded as a vested right." Furthermore, he completely denied that the virtues of mobile warfare could be learned only on the horse and argued that these skills could easily be developed by other means if an effort were made.[55]

The reaction of the Army Council followed the same lines as that of the committee, the civilians generally opposing it, the military members strongly agreeing with it. Both the secretary of state for war, Viscount Halifax, and the deputy undersecretary, R. Paterson, criticized the slow pace established for reducing the number of horses and the expense the government would consequently incur. To them, this was clearly an example of the strong hold that tradition had over the army.[56] On all of these points, the military members of the Army Council disagreed. Led by the same General Montgomery-Massingberd who had headed the Cavalry Committee and who was by now the CIGS, they denied that this was merely a matter of tradition and stressed the great value of the horse in the training of officers. Montgomery-Massingberd summarized their position when he wrote:

> I personally look upon horse exercise and mounted pursuits for the officers as just as important as Physical Training and organized games for the men. I also attach great value to retaining amongst candidates for the Army the class of boy who from his early days has been imbued with a love of the horse and of horse exercises of all sorts. Speaking generally, we draw many of our best officers from this class and we cannot afford to do without him.[57]

Not surprisingly, Montgomery-Massingberd opposed even the modest reductions proposed in the report. His position appeared quite firm initially, but by early December 1935, he softened his views under the pressure of financial considerations. These came in the form of a warning from the permanent undersecretary at the War Office, H. J. Creedy, that the failure to carry out significant reductions in the near future would raise considerable opposition in both the Treasury and the House of Commons.[58] The wrangling as to the exact amount of the reductions continued, however, and it was not until May 1936, a full thirteen months after the report was issued, that a final decision was made. Unfortunately, this decision only called for a reduction, not the abolition, of the number of horses in mechanized units.[59] The actual order implementing this decision was not issued until Novembver 1936 and reductions themselves were not to take place until 1 April 1937 "or as soon after that date as the specified conditions in each case are fulfilled."[60]

Nonetheless, the decision aroused a storm of protest. It prompted General Sir Edmund Ironside to write that it appeared "to be causing a widespread feeling amongst officers of the ex-mounted arms that they are being harshly treated." He continued to say that "I have, moreover, come into contact with many parents who had contemplated sending their sons into the army and who have now become 'anti-army' for these reasons."[61] Reactions such as these did not halt the inevitable, but it is noteworthy that the complete elimination of horses from mechanized units did not occur until shortly after war broke out in 1939.[62]

Instances such as this defense of the horse as a valuable aid in officer's training, together with the more simplistic professions of faith regarding the value of cavalry noted earlier, have long provided ammunition for the charge that the British inability to counter the blitzkrieg was the product of social conservatism within the officer corps. Yet, as the foregoing examination has shown, such a conclusion is not warranted by the facts. By the interwar period, the officer corps was overwhelmingly dominated by the middle class, cavalry officers were increasingly neglectd as general-officer material, and professionalism, while in need of improvement, was neither dead nor confined to the tank advocates. Most important, the belief that the army was blinded to the requirements of modern war by affection for the horse and innately hostile to the tank simply is not true. The fact is, army opinion was not so rigid as it commonly assumed and more restricted in focus. The recommendations of the Cavalry Committee in 1926 and the attitude the Third Hussars nine years later are ample proof of this. Indeed, one of the leading tank advocates has recalled that, while assigned to the War Office during the late 1920s, he served with several cavalry officers who were quite progressive on the tank question.[63] The Scots Greys incident was an exception to the norm, not an example of it. The only time the army displayed any hostility over the loss of horses was in the question of mounts for their personal use. The "Report of the Committee on the Value of the Horse in the Training of Officers" and subsequent events show the army was extremely rigid on this question. But it would be wrong to interpret this hostility as part of a broadly based opposition to mechanization in general. As Montgomery-Massingberd's views on this and the Cavalry Committee report reveal, those officers directly involved in these decisions made a clear distinction between the use of horses for training and their ability in modern war. There is surely considerable truth in the charge by one of the leading tank advocates that this practice ran the danger of retarding the army's mental conversion to mechanization, but there is no indication that it slowed the material transition. Nor was the army blindly hostile to tanks. Those who sought to abolish them completely were very few and isolated, and there is no indication that they had any influence whatsoever. Most critics were deeply concerned with the limitations of tanks, and, while they displayed little imagination regarding the potential

of armor, many of their criticisms were valid for the vehicles of the time. In short, while the army was not so enthusiastic about armor as the tank advocates, claims that mechanization took place only against the "bitter rearguard opposition of the traditionalists" simply cannot be substantiated.

This leaves only Correlli Barnett's assertion that social conservatism spawned a degree of regimental provincialism that severely inhibited the practice of close cooperation among the combat arms. Based as it is on an inaccurate assumption regarding the officer corps, this view is automatically suspect. But, more important, it ignores Haig's own emphasis on cooperation among the combat arms that is found in his dispatches and the examples of such cooperation in the closing stages of World War I. The great British offensive of August 8, 1918 was, as the following chapter will show, a brilliantly conceived and executed combined arms operation. If the British failed to repeat this success in the early stages of World War II, rural provincialism could scarcely have been the reason.

The Effect of Government Policy

Although most historians have agreed that the military policies of the British government during the interwar years slowed the growth of the army's tank forces, they have disagreed on the nature and extent of that effect. Many, such as Barnett and Trythall, have seen it as complementary to the influence of social conservatism; the army did not want to mechanize and the absence of government pressure or incentives gave it a ready excuse not to do so. Both of these historians have particularly stressed the impact of "draconic budget restrictions" on military spending, although Barnett has also asserted that the government enhanced the regimental provincialism of the officer corps by limiting the army's peacetime role to that of an "imperial gendarmerie" in which new ideas were unnecessary.[64] Others, however, have minimized or ignored any conservative influence within the army and placed the blame for the army's lack of success squarely on the government's shoulders. This thesis runs through much of the *History of the Second World War, United Kingdom Military Series,* the British Official History of the War. In his account of the 1940 French campaign, L. F. Ellis has written in general terms that "military failure in 1940 was directly due to the governmental policy of the pre-war years."[65] More specifically Major General I. S. O. Playfair has described the early failures in North Africa as follows:

> The fundamental cause of the troubles of the British in the armoured field was that between the wars they had allowed research and experiment on their own invention the tank, and also its manufacture, to dwindle to almost nothing. The main reason for this state of affairs were

> financial stringency and a policy which for many years assumed that a major war was not to be expected.[66]

Moreover, according to Playfair, this problem was heightened at the beginning of rearmament because the army was not given a "clear idea of what sort of war they might have to fight" and consequently felt compelled to develop a number of designs to fit any contingency. The result was delays in production and added mechanical difficulties with some models of tanks.[67] Nor are army historians the only ones who have made this claim. D. C. Watt, professor of international history at the University of London, has asserted that responsibility for the state of Great Britain's armored forces in 1939 lies with

> the general determination of the civilian strategists not to allow the generals to design a continental expeditionary force, a determination which largely ruled out tank expansion since tanks were thought rather ill-designed for the colonial wars which were all the army were allowed to plan for before the winter of 1938.[68]

Another noted British military historian, Brian Bond, has echoed similar sentiments. Based on his own investigation of the material inadequacies of Britain's armored forces and what he describes as lack of "a coherent doctrine of armored warfare" in 1939, he states that he "fully endorses" the conclusion that these were primarily the result of British government policy, to which Watt and others have referred.[69]

Certainly, there is more truth to these arguments than there is to the "social" interpretation. The army's status within Great Britain's overall military posture has always been far more precarious than that of the other services. The Royal Navy and, after its creation, the Royal Air Force have been recognized as fundamental to the defense of the British Isles, for they were the first line of defense against any direct attack. The army, on the other hand, occupied a secondary role in defending the home islands, and its strength has depended primarily on Great Britain's overseas commitments, fluctuating widely as they increased or decreased. Furthermore, the army has had to face the problem—common to all armies—of relying solely on government funding for the design and production of virtually all its equipment. Navies and air forces can rely on maritime and aviation industries respectively for much of their technology and productive capacity, but armies do not have this advantage; most of their equipment simply lacks civilian counterparts that can be of use.[70] As a result, unless Great Britain has had a definite overseas commitment against a powerful enemy, the maintenance of a strong, well-equipped army has been viewed as a luxury with which the nation could easily dispense.

Unfortunately, as Watt, Bond, and others have emphasized, this was exactly the kind of far-reaching commitment that Great Britain sought to

avoid during the interwar years. Widespread public revulsion against the losses of World War I were undoubtedly a significant element in this desire, but political differences with France and, most particularly, financial considerations were foremost in the minds of those who had to make the decisions. The Liberal and Labour parties continued to adhere to their ideological antimilitarism, and the Conservatives, who dominated the politics of most of the period, loyally reflected the views of the business community, which held that national prosperity through the promotion of business was the first task of any government and was incompatible with heavy military spending.[71] Both foreign and defense policies were viewed in this light, and, more often than not, the army bore the brunt of these decisions. By 1922, army expenditures were only five percent of the 1918 figure, indicating the rapidity with which England demobilized, and they were cut even further in the late 1920s and early 1930s; indeed, total defense spending throughout the interwar period was less than the prewar period when the declining value of the pound is taken into account.[72]

To justify these reduced expenditures, the government first adopted what was known as the Ten-Year Rule. Approved in 1919 at the urging of Winston Churchill, it stated "that it should be assumed for the purpose of framing the estimates of the fighting services, that at any given date there will be no major war for ten years" and that no expeditionary force would be required for this purpose.[73] Protests from senior military officers that such a policy would make it impossible to maintain the army on an effective footing fell on deaf ears. The financial limitations that followed were disastrous. Supplies were not replaced, equipment was not modernized, and the military establishment was even allowed to drop below the minumum requirements set when the rule was established.[74] Realistic strategic planning became impossible: given the rapid changes in technology, particularly in the field of aviation, it was impossible to foresee what weapons might be available in ten years or to formulate a valid plan predicated on their use. Yet, to those outside the armed forces, there seemed to be little reason for concern. The end of the Ruhr Crisis in 1923 and the adoption of the Dawes Plan the following year marked the beginning of a period of international tranquility and economic recovery. The spirit of Locarno made calls for increased military spending appear anachronistic at best. Few people concerned themselves with military policy as it affected either Great Britain's international position or the internal state of the armed forces. Despite constant efforts by the services, the Ten-Year Rule was not recinded until 1932, when the Manchurian crisis forced the government to change its thinking. This made little difference in practice, however, for the government made it clear that Great Britain's economic position still precluded any increases in military spending, cancellation of the Ten-Year Rule notwithstanding.[75]

The rising Nazi menace in the mid-1930s brought the first efforts at

rearmament, but financial considerations remained dominant, especially to the detriment of the army. In 1936, one government advisor urged that priority be given to the air force over the army because the latter was in such a bad state that any attempt rapidly to reequip and strengthen the regular and reserve forces would only produce economic dislocation by draining skilled labor from the civilian economy and creating unattainable demands on the machine-tool industry.[76] Characteristically, the Baldwin government compromised by approving the modernization of the regular army only. In 1937, however, army modernization was unequivocably sidetracked when the new Chamberlain government adopted a policy of "limited liability." In essence, this stipulated that, in the event of a major war on the Continent, Great Britain's strategic priorities should be: first, defense of the home islands against direct attack; second, maintenance of open seaways; third, defense of overseas territories; and fourth, if and when the other three had been accomplished, cooperation with allies in the defense of their territories.[77] Quite explicitly omitted was the intention of building a massive army on the scale of in World War I. It was, in short, a return to the "blue water" strategy that Great Britain had followed in the eighteenth and early nineteenth centuries.

The appeal of the policy of limited liability to the Chamberlain government rested on three considerations. In the first place, it would resolve continued Treasury warnings that a massive rearmament program would wreak havoc with Great Britain's economy.[78] In the second place, it would eliminate the recurring nightmare of many Britons of a repetition of the slaughter of the last war. And, finally, it was argued by many that this was a militarily sound response to the growing tensions and threat of war in Europe. Since the early 1930s, there had been a growing school of thought in Great Britain advocating a return to the "blue water" strategy. Headed by none other than Liddell Hart, who, in 1932 wrote a book entitled *The British Way in Warfare,* this school argued that the "blue water" strategy not only had been highly successful in the past—bringing about the defeat of Napoleon, for instance—but also was an ideal method by which England could apply its strength against a Continental enemy's weakness. By utilizing its command of the sea, Great Britain could prevent an invasion of the home island, protect its lines of communications and overseas territories, wage an effective economic war against the enemy, and strike directly at vulnerable points in the enemy's defenses.[79] By 1936, Liddell Hart was also paralleling some of the Treasury's earlier arguments, claiming that the army was incapable of playing more than a minor role in a Continental war because of its poor state of training and equipment. It was impossible, he concluded, to field an expeditionary force in the near future that would have an effect commensurate with the expenditure, effort, or risk involved.[80]

The government's adoption of the policy of "limitied liability" meant rearmament efforts were henceforth devoted almost entirely to the navy and air force while the army's position was so reduced that the Territorial Army (that is, the army reserve forces) was assigned the mission of manning the antiaircraft defenses in the British Isles rather than reinforcing the regular army if and when it was sent abroad. But for the government, concerned as it was with the economy, this was a small price to pay. Rearmament proceeded without serious dislocation to the civilian economy, and military orders were not even given priority over civilian orders until March 1938.

As long as Great Britain felt sure that the Germans could not overrun France, they saw no reason to doubt the wisdom of this strategy. The elimination of Czechoslovakia as a military power in September 1938, however, meant that France might have to bear the full brunt of a German attack, and under these circumstances, the ability of the French to maintain their positions could not be assured without question. As a result, the British government began to revise its thinking about a Continental commitment at the end of 1938. Early in February 1939, the government agreed that part of the army should be equipped to fight a major power, but before an exact determination could be made as to how this could be done, the policy of limited liability was dropped altogether.[81] Great Britain was now fully committed to supporting France on the land as well as on the sea and in the air; only six months remained before war would break out.

The wisdom of Great Britain's foreign and strategic policy during the interwar years has been as heatedly debated since then as it was at the time, and there can be little purpose in shedding more ink on the subject here. However these policies may be justified, there is little doubt that their impact on the army was severe. Throughout most of the interwar years, the army was starved for new equipment, and the last-minute attempts to rectify this situation while simultaneously undergoing rapid expansion could not possibly have met with the complete success. There is also no doubt that the budgetary and imperial restrictions on the army affected the development of new models of tanks. Until 1936, the army placed its greatest emphasis on the design and production of light tanks, not only because they were cheaper, but also because they were suitable for colonial wars. Consequently, of the seven new tanks projects undertake between 1924 and 1934, six were medium or light tanks, and the seventh, a heavy tank, saw only one pilot model produced.[82] Furthermore, many of the tanks actually produced during the interwar years were upgraded in classification. The Mark I medium tank, the workhorse of the Royal Tank Corps between the wars, was originally designed as a light tank, and the Carden-Lloyd light tank was originally a machine-gun carrier. Concentrated efforts to develop heavier tanks did not begin until after 1937, but by then the resources for designing and producing such vehicles had been largely dissipated by almost twenty years of neglect, and the army experienced considerable technical difficulties and delays in restoring them. When the war broke out, the army possesssed only sixty heavy (infantry) tanks of a total requirement of just over 1,600.[83]

Despite this impressive array of evidence, however, the argument that British government policy was primarily responsible for the army's early defeats has severe limitations. The first is that it fails to explain why the ratio of armored to infantry divisions within the army decreased as England moved toward a definite Continental commitment in late 1938 and early 1939. Logically, if the army's role as an imperial police force had restricted the expansion of its tank forces before 1938, this shift in policy should have produced an increase in the ratio of armor to infantry. But it did not. At the height of the policy of limited liability in mid-1938, the British regular army consisted of one armored division (in the process of being formed) and four infantry divisions, for a ratio of one to four; in February 1939, as the government was moving toward a Continental commitment, it approved a War Office proposal for the expansion of the army, including reserves, to three armored and nineteen infantry divisions, for a ratio of one to six; and when conscription was declared the following month, the army's size was increased to four armored and twenty-eight infantry divisions, for a ration of one to seven.[84] Clearly, the size of Great Britain's tank forces was determined by more than simply the army's role in the nation's defense policy.

Much the same is true for the multiplicity of tank designs during rearmament. The vast extent of the British Empire and the shifting emphasis toward a Continental commitment made it impossible for the army to determine the conditions under which it would have to deploy its main strength, but the successive tank programs it proposed between 1936 and 1939 do not indicate this caused any confusion regarding the types of tanks required. Rather, these programs indicate the army had a very clear perception of the various missions tank forces should perform and a firm belief that these missions (and not the theater in which they would take place) required different types of tanks.[85] The war proved many of these expectations wrong, but that is very different from being confused. Finally, British government policy does not explain why its armored forces continued to exhibit a marked inferiority to the Germans until the third year of the war, and why they were never able to match the early successes of the Panzers even after that. The late start of British rearmament partially explains the initial failure, as many units had to be deployed without adequate training in the handling of their new equipment, but this argument cannot be extended too far. The fact that the army continued to have difficulties successfully employing armor as late as Operation Crusader in November and December 1941, when it had two years of war and five major operations behind it, indicates not so much a lack of training as it does the wrong kind of training. In short, while prewar government policy produced severe equipment shortages and led to a serious decline in the design and production facilities for the heavier classes of tanks—both of which admittedly had an adverse effect on the performance of the tank forces early in the war—it was by no means the only or even most important factor determining the size, composition, and actual performance of those during the opening campaigns of World War II.

The Search for Responsibility

The foregoing analysis demands that a new framework be established to explain the failure of Great Britain's armored forces in the opening campaigns of World War II. Such a framework can be found in the strategic and tactical doctrines of the British Army. These doctrines were the product of a long period of historical evolution, originating in the nineteenth century, and existed totally apart from any social or political influences. The challenge of the new theory of armored warfare proposed by Fuller and Liddell Hart was not so much that it called for the replacement of horses by tanks, but that it attacked these doctrines and the assumptions on which they were founded. The failure of the British Army was not so much that it was anti-tank or that it was prevented from building a powerful tank arm, before the war began; rather, it was that it refused to change its basic strategic doctrines to maximize the potential of a new weapon of war.

Notes

1. B. H. Liddell Hart, *Memoirs,* 2 vols. (London: Cassell, 1965), 1:77.
2. Anthony Trythall, *"Boney" Fuller* (New Brunswick, N.J.: Rutgers University Press, 1977), p. 159.
3. Kenneth Macksey, *Armoured Crusader* (London: Hutchinson, 1967), p. 71.
4. Correlli Barnett, *The Desert Generals* (New York: The Viking Press, 1960), pp. 98–103.
5. Correlli Barnett, *Britain and Her Army* (London: Allan Lane, The Penguin Press, 1970), p. 420.
6. P. E. Razzell, "Social Origins of Officers in the Indian and British Home Army," *British Journal of Sociology* 14 (September 1962): 353, Tables 7 and 8.
7. Donald C. Watt, *Too Serious a Business* (Berkeley, Calif.: University of California Press, 1975), p. 71.
8. Great Britain, War Office, *The Half Yearly Army List for the Period Ending 31 December 1922* (London: HMSO, 1923), pp. 6–163; Great Britain, War Office, *The Monthly Army List for August 1932* (London: HMSO, 1932), pp. 150–58; Great Britain, War Office, *The Half Yearly Army List for the Period Ending 31 December 1938* (London: HMSO, 1939), pp. 9–169. For the purposes of this analysis, officers from noncombat branches were excluded.
9. Ibid.
10. Great Britain, War Office, *War Office List* (London: HMSO, 1928), p. 4. In October and November 1937, the Army Council was reorganized, adding the director general of the territorial army (who previously had been under the AG) and replacing the MGO with the director general of munitions production, who was to control weapons design and production for all three services.

11. These appointments were actually made to only twenty-three men as one, General Sir A. A. Montgomery-Massingberd, held two positions, AG and CIGS. This information is compiled from the *Army Quarterly* from 1920 to 1939, which annually published the names of those serving on the Army Council, "Army Notes" published in each issue of the *Journal of the Royal United Service Institution* from 1923 to 1939, which contained notices of promotions and appointments, and the *Half Yearly Army Lists* for 1923 and 1938, which provide the branch of service for all officers on active duty.

12. General Sir John Burnett-Stuart, "Extract from Report on the Calvalry Brigade's Training," an untitled memorandum dated September 1935. Lindsay Papers, States House, Medmenham, Buckinghamshire, England.

13. C. Bonham-Carter, "The Royal Military College, Sandhurst," *Army Quarterly* 3 (October 1921): 105.

14. For examples of this aspect of British Army life by officers who served in it at the time see Colonel R. MacLeod, "An Artillery Officer in the First World War," an unpublished manuscript in the Fuller Papers, Box 17, Special Collection Department, Rutgers University Library, New Brunswick, N.J.; and Great Britain, War Office, *Report on Army Education* (London: HMSO, 1902), p. 131, Appendix 40.

15. *Report on Army Education,* p. 28.

16. Quoted in Brian Bond, *The Victorian Army and the Staff College* (London: Eyre McThuen, 1972), pp. 181–82.

17. Ibid., pp. 367–68. War Office File (hereafter cited as WO) 32/4840, "Report on Higher Education for War and Reorganization of the Senior Division of the Staff College," Minute 12A.

18. Bond, *The Victorian Army and the Staff College,* pp. 196–98, 244–73. This compares with an average of twenty-nine officers who applied for the fifteen vacancies open annually in the 1860s.

19. Ibid., p. 30.

20. Barnett, *Britain and Her Army,* p. 347. A peak of 450,000 men was raised during the Boer War, approximately one-half of the German peacetime strength at that time.

21. Captain Wilfred Miles, *Military Operations France and Belgium: 1916, July 2 to the End of the Somme Battles,* in *History of the Great War Based on Official Documents,* Brigadier General Sir James E. Edmonds, gen. ed. (London: MacMillan and Co., Ltd., 1938), p. 235.

22. Barnett, *Britain and Her Army,* pp. 410–11.

23. Anonymous, "Editorial," *Army Quarterly* 9 (October 1924): 2.

24. A. R. Godwin-Austin, *The Staff and the Staff College* (London: Constable and Co., Ltd., 1927), pp. 291, 296.

25. WO 32/4840, "Report on Higher Education for War and Reorganization of the Senior Division of the Staff College," Minute 12A. For an extended treatment of this subject, see B. Bond, *British Military Policy between the Two World Wars* (Oxford: Clarendon Press, 1980), pp. 35–71.

26. Douglas Haig, *Sir Douglas Haig's Dispatches,* ed. J. H. Boraston (London: J. M. Dent and Sons, Ltd., 1920), dispatch of 21 March 1919, p. 327. For an earlier example of Haig's praise of the cavalry, see his dispatch of 21 December 1918, p. 301.

27. This comment has been quoted, for example, in Robert Graves and Alan Hodge, *The Long Week-End: A Social History of Great Britain, 1918–1939* (New York: W. W. Norton and Co., 1963), p. 196; Liddell Hart, *Memoirs,* 1:100; and Peter Dennis, *Decision by Default: Peacetime Conscription and British Defense, 1919–1939* (Durham, N.C.: Duke University Press, 1972), p. 15.

28. Anonymous, "Parliamentary Notes," *Army Quarterly* 2 (July 1921): 415.

29. Anonymous "A Horse! A Horse! My Kingdom for a Horse!," *Army Quarterly* 28 (April 1934): 150–53.

30. Anonymous, "The British Campaign in the West, Pt. II," *Army Quarterly* 6 (April 1923): 44–56.

31. A. P. Wavell, "The Strategy of the Campaign of the Egyptian Expeditionary Force," *Army Quarterly* 3 (January 1922): 235–48. For other examples of this same view, see A. G. Wauchope, "The Destruction of the VIth Turkish Army," *Journal of the Royal United Service Institution* 54 (August 1919): 436–59; and Harold Charrington, letter to the *Daily Telegraph,* 28 March 1933.

32. Charrington, letters to the *Daily Telegraph,* 28 March and 1 April 1933.
33. F. A. S. Clarke, "Some Further Problems of Mechanical War-Fare," *Army Quarterly* 14 (July 1927): 384.
34. Victor Germains, *The Mechanization of War* (London: Praed and Co., Ltd., 1928), p. 40.
35. Ibid., p. 48.
36. Haig, *Sir Douglas Haig's Dispatches,* dispatch of 21 December 1918, p. 302.
37. Germains, *The Mechanization of War,* pp. 101–7, 166–67, 234–43.
38. WO 32/2841, "Cavalry Committee—Interim Report," Minute 3A, pp. 26–28. The conclusions of the committee were presented in this report and in WO 32/2842, "Cavalry Committee—Final Report."
39. WO 32/2841, "Calvary Committee—Interim Report," p. 4.
40. Ibid., pp. 4–5, 10–11, 24.
41. Ibid., pp. 5–7, 10–11, 15–16. The committee recommended this method of reduction rather than a reduction in the number of regiments so as not to disrupt the army's organization tables.
42. Ibid., pp. 5–7
43. *Daily Telegraph,* 6 September 1928.
44. W.O. 32/2831 Minute 30.
45. Commandant of the Equitation School to Montgomery-Massingberd, 19 December 1926, and B. Fisher to Montgomery-Massingberd, 9 December 1926, File 155, Montgomery-Massingberd Papers, Kings's College, University of London.
46. WO 32/4195, "Army Council Informal Meetings," Meeting Number 7, 22 October 1937, p. 8. See also WO 32/4633, Minute 24B.
47. WO 32/4633, "The Suggested Retention of the Scots Greys as Horsed Cavalry, 1936–37," Mintues 15A, C; 16A, C, E, G, L, M; 17A, B, F, H, K, M are the file copies of the letters sent.
48. WO 32/4195, Minutes of meeting NO. 9, p. 6; and WO 32/4633, minutes 23A, 23B, 25A.
49. Grubbe, "Report of the Mechanization of the 3rd King's Own Hussars," 26 September 1935, Minute 13A in WO 32/2847, "Introduction of the Mobile Division," 1934–36.
50. Ibid.
51. General Sir Frederick Pile, personal interview with the author at the Cavalry Club, London, on August 6, 1971.
52. Hobart to Liddell Hart, 31 May 1938, Liddell Hart Papers, States House, Medmanham, Marlow, Buckinghamshire.
53. WO 32/4005, "Report of the Committee on the Value of the Horse in the Training of Officers," p. 4.
54. Ibid., pp. 5, 8–9.
55. Ibid., pp. 12–13.
56. Ibid., Minute 5A.
57. Ibid., Minute 7, a memorandum by Montgomery-Massingberd dated 14 October 1935.
58. Ibid., Minutes 39 and 41.
59. Ibid., Minute 47A. The letter so ordering was dated May 1936.
60. Ibid., Minute 51A. The order was dated 20 November 1936.
61. Ibid., Minute 78A.
62. Ibid., Minute 101A.
63. Interview with General Pile.
64. The quote is from Barnett, *Britain and Her Army,* p. 415. See also *The Desert Generals,* p. 100, and Trythall, *"Booney" Fuller,* p. 159.
65. Major L[ionel] F. Ellis, *The War in France and Flanders,* 1939–40, in *History of the Second World War, United Kingdom Military Series,* Sir J[ames] R. M. Butler, gen. ed. (London: HMSO, 1953), p. 315.
66. Major General I[an] S. O. Playfair, *The Mediterranean and Middle East,* vol. 2: *The Germans Come to the Help of Their Ally,* in *History of the Second World War, United Kingdom Military Series,* Sir J[ames] R. M. Butler, gen. ed (London: HMSO, 1956), p. 174.
67. Ibid., pp. 174–75.
68. Watt, *Too Serious a Business,* p. 81.
69. Bond, *British Military Policy between the Two World Wars,* pp. 180 and 189. Bond also gives

some weight to conservatism within the army in producing this situation, but emphasizes his belief that this should be viewed as a secondary cause. Ibid., pp. 187–88.

70. M. M. Postan, D. Hay, and J. D. Scott, *The Design and Development of Weapons,* in *History of the Second World War, United Kingdom Civil Series,* Sir Keith Hancock, gen. ed. (London: HMSO, 1964), pp. 238–39.

71. R. P. Shay, *British Rearmament in the Thirties* (Princeton, N.J.: Princeton University Press, 1977), pp. 14–17.

72. Robin Higham, *Armed Forces in Peacetime: Britain 1918–1940, A Case Study* (Hamden Conn.: Anchor Books, 1962), Appendix 1, pp. 326–27; Dennis, *Decision by Default,* p. 13.

73. Hay, Poston, and Scott, *The Design and Development of Weapons,* p. 11.

74. Shay, *British Rearmament in the Thirties,* p. 20.

75. Ibid., p. 23.

76. Ibid., p. 80. The advisor was Lord Wier, technical advisor to the Defense Requirements Subcommittee of the CID.

77. Ibid., p. 168. See Dennis, *Decision by Default,* pp. 100–114, and Bond, *British Military Policy between the Two World Wars,* pp. 214–86, for a complete account of the policy of "limited liability."

78. Shay, *British Rearmament in the Thirties,* p. 167.

79. B. H. Liddell Hart, *The British Way in Warfare* (London: Faber and Faber, Ltd., 1932), pp. 12–56.

80. *The Times,* November 2, 1936.

81. Dennis, *Decision by Default,* pp. 179–80.

82. Hay, Postan, and Scott, *The Design and Development of Weapons,* pp. 306–7.

83. Sir J[ames] R. M. Butler, *Grand Stragegy,* vol 2: *September 1939–June 1941,* in *History of the Second World War, United Kingdom Military Series,* Sir J[ames] R. M. Butler, gen. ed. (London: HMSO, 1957), p. 28.

84. See below, p. 219.

85. See below, pp. 191–92, 199–201.

• 2 •

The Tradition: The Strategy of Attrition

THE victory bells that pealed throughout England on the morning of November 11, 1918 signaled more than just the end of the war; to the generals who had led the British Army to this moment of victory, the bells also signaled that the strategic principles that had guided their conduct of the war were sound and the faith placed in them justified. It had not been an easy course to follow. For almost four years they had been buffeted by charges from their own political leaders that they were wasting the lives of the nation's young men in massive, senseless offensives; that the road to victory lay not in the fields of France or Belgium, but in the Dardenelles, Balkans, or Italian Alps. The generals fought long and hard against these charges. In the end they prevailed, not because they converted their opponents, but because the government could find no alternative to their leadership.

Yet this failure to convert their opponents did not shake the generals' faith in their conception of war, and their successors of the interwar period were equally rigid. They recognized that advances in technology produced changes in equipment requiring modifications in the way that strategic principles were applied. But, they insisted, these principles remained constant. The evolution of the tank during the war was an excellent case in point. Although the weapon itself was a radicasl innovation, its role as conceived both by the men who developed it and by the British High Command was strictly within the confines of the army's traditional strategic doctrines. This view continued to shape the evolution both of the British Army and the Royal Tank Corps during the interwar period.

Prewar Military Doctrine

The British Army's strategy in World War I was the product of a century-long search to discover the secret of Napoleon Bonaparte's battlefield success. The emperor's victories had inspired innumerable studies filled with reverence for his achievements and claims of having penetrated the innermost recesses of his mind. Most of these works soon passed into well-deserved oblivion. But by the late nineteenth century, a definite model of war rooted in the Napoleonic experience emerged, and became the basis for the strategic and tactical doctrines of most of the European powers. According to this model, war was a gladiatorial contest in which the sole legitimate objective was the destruction of the enemy's armed forces. It held that such destruction could only be achieved in decisive battles and insisted that the only legitimate actions were those designed to help achieve that end. The act of battle, in broad terms, was divided into four distinct phases, the first and last of which were highly mobile, the middle two, relatively static. In the first, actually a preliminary stage, the contending forces maneuvered against each other to seize control of key geographic features and place the enemy at a disadvantage. In the second, they sought to immobilize or weaken each other through a process of attrition—that is, by conducting "holding" or "wearing-out" attacks—while massing their own reserves. When one side had gained the upper hand, the third stage opened. This was the "decisive attack," a massive assault launched by the superior army with all available resources against a vital sector in the enemy's position and designed to rupture his line and force him into headlong retreat. When it succeeded, the fourth or pursuit phase opened, during which the remnants of the enemy's forces were destroyed. The keys to the successful execution of such a plan were held to be unrelenting offensive pressure both before and during the battle, the application of maximum force in the decisive attack, and an unswerving determination by both commanders and troops to conquer at any cost.

It is generally agreed that the origins of this model are contained in the writings of Karl von Clausewitz, the great Prussian military theorist of the early nineteenth century. This is true—but only in a limited sense. Clausewitz was a product of the two dominant forces of his age, the Napoleonic Wars and the German philosophical spirit of fundamental causes epitomized by Immanuel Kant. In his classic treatise, *On War,* he brought these two themes together in a penetrating if sometimes confusing synthesis. From his wartime experiences, Clausewitz observed firsthand how the French Revolution had drastically changed the limited nature of eighteenth-century wars, which thereby enabled first the French and then their opponents to win decisive battlefield victories. From German philosophers, he learned to search for the absolute or "regulative idea of things." To this end, he established the concept of "absolute war," not "total war" as experienced in the twentieth century, although there are superficial similarities, but a philosophical abstraction by which war could be placed in a vacuum for study and analysis. Once the essence of war had been ascertained, he believed, it could then be applied to concrete situations, producing a full understanding of both the constant forces operating in war and the variations produced by specific conditions.[1]

Unfortunately for Clausewitz, this concept of "absolute war" did not impress his readers as he intended. The limitations of reality that he imposed on his theoretical model of war were not understood, and his concept of "absolute war" was viewed as an astute analysis of the secrets of Napoleonic warfare, a blueprint for future success. The result was the adoption of the model described above. To the military leaders of Europe, war now became a purely military struggle in which numbers and determination were the sole prerequisites for success. Part of the blame for this distortion undoubtedly lay with Clausewitz—his writing tended to blur the distinctions between his concepts of "absolute" and "real" war—but most of the fault was due to readers who were blinded by the superficialities of Napoleon's success. As Michael Howard has remarked in discussing the broader implications of the emperor's career: "The fact remained that Napoleon lived by the sword and he perished by the sword. The study of swordsmanship thus acquired a heightened significance in the eyes of posterity."[2]

Even the most cursory reading of *On War* reveals statements that can form the basis for this distortion of Clausewitz's ideas. Regarding the objective of war, for example, he wrote: "The destruction of the enemy's armed forces, amongst all the objectives which can be pursued in War, appears always as the one which overrules all others."[3] On the means to this end, he was even more assertive: "Because the essence of War is conflict, and the battle is the conflict of the main Armies, it is always to be regarded as the real centre of gravity of the War, and therefore its distinguishing character is, that unlike all other encounters, it is arranged for, and undertaken with the sole purpose of obtaining a decisive victory."[4] And again: "The destruction of the enemy's military force is to be sought for principally by great battles, and their results; and . . . the chief object of great battles must be the destruction of the enemy's military force."[5] And, finally:

> The battle may therefore be regarded as War concentrated, as the centre of the effort of the whole War or campaign. As the sun's rays unite in the focus of the concave mirror in a perfect image, and in the fullness of their heat; [so] to[o] the forces and circumstances of War, unite in a focus in the great battle for one concentrated utmost effort.[6]

As to success in battle, Clausewitz noted first that superiority in numbers "is in tactics, as well as in Strategy, the most general principle of victory," and explained: "An unbiased examination of modern military history leads to the conviction that the *superiority in numbers becomes every day more decisive;* the principle of assembling the greatest possible numbers for a decisive battle may therefore be regarded as more important than ever."[7] He closely linked this superiority to psychological factors. Asserting that "the condition of the mind has always [been] the most decisive influence on the forces employed in war,"[8] he concluded that, since the confusion of battle prevents an accurate assessment of relative casualties, the chief reason for the collapse of morale is a commander's inability to parry enemy thrusts with fresh troops. As a result, "the loss of moral force may be measured by the reserves that have been used."[9] As an experienced soldier, Clausewitz was sufficiently realistic to see that such a concept of war could well lead to even higher casualties than in the past, but denied the existence of a viable alternative:

> The battle is the bloodiest way of solution. True, it is not merely reciprocal slaughter, and its effect is more a killing of the enemy's courage than of the enemy's soldiers . . . but still blood is always its price, and slaughter its character as well as name. . . .
>
> Thus, then, Statesmen and Generals have at all times endeavoured to avoid the decisive battle, seeking either to attain their aim without it, or dropping that aim unperceived. Writers on history and theory have then busied themselves to discover in some other feature in these campaigns not only an equivalent for the decision by battle which has been avoided, but even a higher art. . . .
>
> Contempory history has destroyed this illusion, but no one can guarantee that it will not sooner or later reproduce itself. . . .
>
> Let us not hear of Generals who conquer without bloodshed. If bloody slaughter is a horrible sight, then that is a ground for paying more respect to War, but not for making the sword we bear blunter and blunter by degrees from feelings of humanity, until some one steps in with one that is sharp and lops off the arm from our body.[10]

Much of the interwar debate over British strategy would center on this very point: whether or not armored forces could achieve victory without paying such a price.

Although Clausewitz placed many qualifications on these statements and the body of this work is far more sophisticated than they suggest, the fact remains that this is the essence of his bequest to generations of European soldiers. While *On War* was widely read throughout the Prussian Army in the decades following his death in 1831, it was only with the Prussian victories over France in 1870–71 that his reputation spread to the rest of Europe. These victories affected a revolution in military affairs. Numerous aspects of the Prussian military system were adopted by other states, including the institution of the general staff, the system of annual field maneuvers, elements of strategic and tactical doctrine, terminology, uniform patterns, and, perhaps most important of all, conscription.[11] Short-term, universal service became the rule, with quantity replacing quality as the main criterion for judging military strength. Germany led the way: between 1870 and 1914, its regular army increased from 400,000 to 850,000, and was backed by a reserve of five million.[12] With a smaller population, France was forced to conscript every man possible, and although it was able to match the German forces in 1914 it had to raid the armament industry to do so.[13]

This emphasis on mass was reflected in the development of battlefield tactics. In the decade after the battle of Sedan, European military doctrine reflected growing respect for the power of rapid-fire weapons and a realization that the key to successful offensive action was dispersion. Tactical manuals asserted that swarms of advancing skirmishers were the best offensive formation, and serious study was made of the problem of reconciling individual initiative and originality with the necessary discipline. By the 1880s, however, the memories of the wars dimmed, and the concept of "building up the firing line" replaced that of dispersion. This concept was essentially the same as that used in the eighteenth century, when troops marched into the attack shoulder to shoulder in rigid formation, though in its revived form much of the former rigidity was lacking. It provided the obvious advantages of increasing the fire power of assaulting troops and enabling commanders to exercise control with greater ease. Furthermore, it was argued, since attacking forces could concentrate their strength at the point of attack while the defenders would have to man their entire line, the former have the advantage of a greater concentration of troops and, therefore, firepower. These changes, first reflected in the French field regulations of 1884, rapidly spread to all the Continental powers.[14] By World War I, the trend was so far advanced that attacking troops often advanced with little more than a yard between them.[15]

By the end of the nineteenth century Clausewitz had imitators and disciples throughout Europe, but few achieved the influence of Ferdinand Foch in France and, later, England. As professor at the Ecole de Guerre from 1895 to 1900 and commandant from 1908 to 1914, his ideas permeated the entire French Army, and the published editions of his lectures became standard reference works.[16] Unfortunately, the errors of distortion in Clausewitz became errors of principle in Foch. The essential difference between the two is that, where absolute war had been a theoretical abstraction to the former, it was a logical and inevitable product of the nation in arms to the latter. In his first important book, *The Principles of War,* Foch admitted that "at such times of general exhaustion, of decreased ambitions and of reduced means," limited wars were possible; but he argued that the German adoption of the nation in arms made such wars impossible in his own day: "It is because all of Europe has returned to this theory of armed nations that we must again take up today the *absolute* conception of war."[17]

In describing the characteristics of this "absolute conception of war," Foch gave paramount importance to the role of battles. He criticized the eighteenth-century conduct of war, holding that "the out-of-date methods, as far as we are concerned, are those of a war without decisive results, with limited purpose, war of maneuver without battles."[18] He concluded:

> Modern war returns to the principle of a *decision by arms,* and no other can be accepted. Instead of blaming the battles of Bonaparte as barbar-

> ous actions it recognizes in them the only efficient means; it also seeks battle in the same manner.[19]

Foch also placed great emphasis on the psychological factors in war. In doing so, he was influenced not only by Clausewitz but also by Colonel Ardant du Picq, a French officer killed in the Franco-Prussian War. Du Picq had become interested in the source of battlefield success during service in Algeria, where he observed that small numbers of French troops consistently defeated far larger numbers of natives. In an effort to discover the reasons for these victories, he studied the ancient wars between the Romans and Germans. His conclusions, published shortly before his death, were that

> in battle, two moral forces, even more than material forces, are in conflict. The stronger conquers . . . with equal or even inferior power of destruction, he will win who is determined to advance, who . . . has moral ascendency.[20]

Foch went well beyond this point in his *Principles of War;* in fact, he made battle essentially a contest of will in which casualties were largely irrelevant. He wrote:

> "A hundred thousand men," says General Cardot "leave ten thousand of their number on the ground and acknowledge defeat: they retreat before the victors who have lost just as many men, if not more. Besides, neither know when the retreat occurs, what their losses are or what the enemy's casualties may be." It is not, therefore, through the material factor of losses, and still less through any comparison of figures, a greater number of casualties, that they give in, renouncing the fight and abandoning to the opponent the ground in dispute.
>
> Ninety thousand defeated men withdraw before ninety thousand victorious men solely because they have had enough, and they have had enough because they no longer believe in victory, because they are demoralized and they have no *moral* resistance left. Which leads Joseph de Maistre to say: "A battle lost is a battle one believes one has lost for a battle is never lost materially. And if battles are lost morally, they must also be won in the same way. So that we can add: "*A battle won is a battle in which one refuses to acknowledge defeat.*"
>
> DeBrach, following Frederick, passes by an old castle in Silesia. Over the door is a coat of arms: two stags with locked horns, and for the motto: *The most obstinate wins.* "That is the truest source of success!" exclaims the famous general.[21]

Foch departed from Clausewitz on only one point: the power of the offensive. The Prussian had been greatly impressed with the power of the defense during his campaigns and in his writing had stressed that it was the stronger form of war. Foch maintained the opposite. Following a survey of the effect of modern weapons on military operations, he declared that "all improvements in firearms add to the strength of the offensive intelligently planned because the attackers, choosing their ground, can concentrate on it so much a greater volume of fire."[22] As a result of Foch's emphasis, belief in the power of the offensive assumed ominous proportions. By the eve of World War I, it was seen not just as a form of maneuver, but as a mystical force capable of solving any strategic or tactical problem. This extreme position was forcefully stated when, shortly after he assumed the post of commander in chief of the French Army in 1911, General Joffre declared:

> The French Army, returning to its tradition, no longer knows any other law than the offensive. . . . All attacks are to be pushed to the extreme . . . to charge the enemy with the bayonet in order to destroy him. . . . This result can only be obtained at the price of bloody sacrifice. Any other conception ought to be rejected as contrary to the very nature of war.[23]

As serious a distortion of Napoleonic warfare as this model was, an even more tragic omen for the future was the growing rigidity in military thought. The period immediately following the Franco-Prussian War had been one of great flux in military theory as the armies of Europe sought to adjust their organizations and doctrines to meet the Prussian challenge. With the general acceptance of this model, however, they concluded that they had found all the answers and that further study and development were unnecessary. Such dogmatic military thinking was certainly not new, but its danger was reinforced at this time by two factors that gave it a unique character: the emphasis on "will" as the key to victory and an increasing appreciation of the role of the general staff. The influence of the former was that the more "will" was stressed, the less possible rational thought became. Be equating victory with desire, any questioning of official doctrine could conceivably undermine confidence and consequently be equated with treason. In the case of Foch, it has been suggested that his strict Jesuit Catholic upbringing contributed to his dogmatism,[24] and, while this may be questioned, there is little doubt that he opposed any deviation from the official doctrine and effectively indoctrinated the French Army in his way of thinking once he became commandant of the Ecole de Guerre.[25] The development of general staffs throughout Europe reinforced this rigidity, as one of their chief objectives was to establish doctrinal uniformity for each entire army. Their goal was to insure that under any set of circumstances a commander could expect approximately the same course of action to be taken by his subordinates. They believed that this was essential because the complexities in maneuvering, supplying, and maintaining communications in the larger armies then coming into existence no longer allowed room for individualists; such men, indeed, could be dangerous to the movement and action of the army as a whole. In its most extreme form—quite common before 1914—this belief led to statements such as that of a British officer who wrote shortly after the war: "The truth is that for a doctrine to be truly effective, it must permeate the mind so as almost to control it. It must in fact exercise the power over the High Command which discipline exercises over the private soldier in moments of the severest tension."[26] The British played virtually no role in the development of this model; in fact, they did not accept it completely until the first decade of the twentieth century. The two most important military writers of the late nineteenth century, General Sir John Frederick Maurice and Colonel G. F. R. Henderson, both realized that rapid-fire weapons were giving increasing advantage to the defense and that massed frontal assaults were becoming an anachronism. Analyzing the Franco-Prussian War, Maurice concluded that the greatest Prussian victories were achieved not by "a brilliant application of offensive strategy but by defensive tactics" and noted that their offensives suffered heavy losses until the final moment of victory, if and when this moment came.[27] Colonel Henderson, whose reputation rested primarily on his studies of the American Civil War, also realized the significance of increased firepower and the development of intrenchments and prophesied that they would play an even greater role in the future. He also concluded that under most circumstances, frontal assaults were far too costly and dangerous, but if they did prove necessary, speed and secrecy rather than mass would be the key to success.[28]

The abandonment of these reservations began with the Haldane reforms in 1906.[29] In addition to creating a general staff and reorganizing the system of reserves, Haldane also undertook the first plans for the possible dispatch of an expeditionary force to the Continent and initiated military discussions with the French. Lacking a firmly established military doctrine for such operations, the British rapidly came under the influence of the French and particularly Ferdinand Foch. The agent responsible for the transfer of these ideas to the British Army was General Sir Henry Wilson, commandant of the Staff College from 1907 to 1910 and a close friend and admirer of Foch. Although recent studies have questioned whether Wilson fully accepted Foch's ideas on the power of the offense, there is little doubt that he believed one of the first duties of the Staff College was to establish a 'School of Thought' which would in time create a sense of uniformity and harmony in the Army as a whole,"[30] and that this meant the introduction of many Continental theories into the British Army.

The consequences of these developments were reflected in the *Field Service Regulations* of 1909, which remained substantially unchanged until after the war began.[31] Although this manual did not emphasize the importance of mass to the same extent as Continental theorists and showed greater awareness of the effect of modern firepower, its basic concept of military operations flowed straight from the Napoleonic model as interpreted by Clausewitz and Foch. It stressed that the objective of all operations was the destruction of the enemy's main forces in battle and, after describing the preliminary movements leading to this, concluded: "*The defeat of the hostile advanced troops is, however, only a means toward the subsequent destruction of the enemy's main force on the battlefield, and this ultimate objective must be held in view.*"[32]

The *Regulations* were equally clear on the conduct of battle. Once contact with the enemy was made, it stated, vigorous offensive action was required to keep him off balance and to pin down and exhaust his reserves while bringing one's own forces together for the decisive blow. In the words of the manual: "*The general principle is that the enemy must be engaged in sufficient strength to pin him to his ground and to wear down his powers of resistance, while the force allotted to the decisive attack must be as strong as possible.*"[33] Throughout this process, the demand for vigorous offensive action was deemed to be of paramount importance. The phrases "decisive blow" and "decisive attack" recur frequently. Readers were told that "decisive success in battle can be gained only by a vigorous offensive."[34] By contrast, the defensive was pictured as a weak, almost unmanly, form of war: it should be "resorted to when and where it is desired to delay a decision [as opposed to] the offensive [which should be used] where it is desired to obtain one."[35]

Foch's emphasis on "will" also found its counterpart in the 1909 *Field Service Regulations.* Claiming "*half-hearted measures never attain success in war and lack of determination is the most fruitful source of defeat,*" it concluded:

> Superior numbers on the battlefield are an undoubted advantage, but skill, better organization, and training, and above all *a firmer determination in all ranks to conquer at any cost are the chief factors of success.*[36]

These were the ideas with which Great Britain and her allies went to war in 1914.

World War I and the Origins of the Tank

The Allied conduct of World War I was at the time and continues to be the subject of bitter controversy.[37] To some the British and French generals were unimaginative, inflexible, and grossly callous in their disregard for the lives of the men under their command. To others, the generals were commanders responsible for a herculean undertaking that could not be completed easily or cheaply, and they assert that charges such as those just noted only revealed a lack of understanding of the nature of modern war and the principles governing its conduct. In their view, the problem was that the size of modern armies and their ability to rapidly shift reserves to threatened sectors by railroad had prolonged the second or attrition phase of their classic model of battle. They remained convinced that the prewar doctrines were correct in every essential, and that the principles contained therein only had to be applied with sufficient determination, perseverence, and vigor to succeed. The fact that this took longer than expected and resulted in more casualties than originally anticipated was certainly no reason to change them. As Field Marshal Haig explained when justifying further offensives against the Germans in the spring of 1917:

> The guiding principles are those which have proved successful in war from time immemorial, viz., the first step must always be to wear down the enemy's power of resistance and to continue to do so until he is so weakened that he will be unable to withstand a decisive blow; then with all one's forces to deliver the decisive blow and finally to reap the fruits of victory.
>
> The enemy has already been weakened appreciably, but time is required to wear down his great numbers of troops. The situation is not yet ripe for the decisive blow. We must therefore continue to wear down the enemy until his power of resistance has been further reduced.[38]

Changes, when they were made, were confined to traditional strategic doctrines. Allied commanders realized, for example, that their inability to rupture the German lines was the result of the firepower of the machine gun and the protection of complex networks of trenches, which gave the defense a degree of power hitherto unknown. The solution, according to prewar thought, was to neutralize these elements with artillery prior to the assault. As a result, when the great artillery bombardments failed to produce the desired effects, the generals resorted to increasing the length and intensity of the bombardments. Thus, at the battle of Hooge in 1915, the British attack was preceded by an artillery preparation of a mere 18,000 rounds; at the Somme the following year, this was increased to 2,000,000; at Arras in 1917, to 2,600,000; at Messines it was again increased, to 3,500,000 rounds; and at Passchendaele (also known as the third Ypres or Flanders campaign), it reached a total of 4,300,000 rounds. This last bombardment lasted for nineteen days, allowing the Germans more than ample time to bring up reserves just out of artillery range. In addition, it destroyed the Flanders' drainage system, turning the battlefield into such a morass of mud when the fall rains came that, by itself, the ground was a formidable obstacle to attackers.[39]

The casualty figures assumed equally grim proportions. The battle of the frontiers cost each side half a million casualties, demonstrating the effectiveness of the machine gun and rapid-fire rifles, but this made little impression on Allied generals. Until well into 1916, the doctrine of "building up the firing line" remained dominant in the British Army as troops advanced with little more than a yard between them. Even when this practice was modified, the vulnerability of advancing infantry still insured a high rate of casualties.[40] The results were devastating to a generation of European manhood. British losses for 1915 totaled 313,000; for 1916, 651,000; and for 1917, 750,000. Both French and German losses were significantly higher.[41]

It was in this situation that the tank was conceived and developed, not as a revolutionary agent to alter radically the conduct of war, but merely as a new weapon to provide the offensive with greater striking power by helping to neutralize the enemy's barbed wire and machine guns.

Credit for the idea of a cross-country armored fighting vehicle—in short, the tank—has been accorded to Lieutenant Colonel (later Major General Sir) Ernest D. Swinton. A Royal Engineer officer attached to the Committee of Imperial Defense in 1914, Swinton was sent to France when the war broke out as the army's official correspondent to cover the fighting for the British press. In this capacity, he visited the front as often as possible to interview the troops regarding their experiences. As the Allies sought to turn the German flank following the battle of the Marne, he became alarmed by what he learned of the infantry's difficulties against the hastily organized German defenses, and he was even more disturbed by thoughts of what would occur when they came up against fully prepared positions. Vaguely, he began to picture "some form of armoured vehicle immune against bullets, which should be capable of destroying machine guns and of ploughing a way through wire."[42] The idea of a self-contained fighting vehicle combining protection and mobility was not, as Swinton was aware, new. The concept went back at least as far as Leonardo da Vinci and only a decade before the war had been the subject of a provocative short story by H. G. Wells.[43] But these earlier vehicles had been wheeled and Swinton realized that such vehicles would quickly become immobilized on modern shell-scarred, trench-lined battlefields. A solution to the problem eluded him until October 19, 1914, when he suddenly recalled a note he had received on the eve of the war describing an American-built tractor on display in Antwerp. In a flash, the idea struck him: by substituting the caterpillar tracks of the tractor for wheels, a vehicle could be constructed capable of crossing all but the most difficult terrain and the largest trenches.[44]

The development of the tank from that October day until it first went into combat almost two years later was marked by considerable confusion and frustration. While in London on official business on October 20, Swinton met with Colonal Maurice Hankey, then secretary to the Committee of Imperial Defense, and T. G. Tulloch, a retired artillery officer. Hankey and Tulloch were both impressed with Swinton's idea and the three agreed to work together to interest higher authorities in the project. After the meeting, Swinton tried to see Lord Kitchner, then secretary of the state for war. As the army's official correspondent in France, he had been promised direct access to Kitchner, but his efforts failed. The next day, he returned to France, and shortly thereafter he presented his idea to the engineer in chief of the British Expeditionary Force. Although the officer gave Swinton a sympathetic hearing, he showed no inclination to follow it through. Somewhat disheartened, Swinton returned to his duties convinced that his efforts had been a total failure.[45]

But the seed planted in Hankey's mind proved far more important than Swinton imagined. During the Christmas holidays, Hankey took the opportunity afforded by a few days' rest to draw up an "appreciation of the war situation as a whole." Known as the "Boxing Day Memorandum," this far-reaching document stressed that the Allies possessed no means of breaking the German lines at that time and suggested a number of solutions for the problem. Among them was a proposal for:

> numbers of large heavy rollers, themselves bullet proof, propelled from behind by motor engines, geared very low, the driving wheels fitted with a "caterpillar" driving gear to grip the ground, the driver's seat armoured, and with a Maxim gun fitted. The object of this device would be to roll down barbed wire by sheer weight, to give some cover to the men creeping up behind, and to support the advance with machine gun fire.[46]

Hankey's memorandum was passed around the War Council[47] as a matter of course, and it struck a responsive chord in Winston Churchill, then first lord of the Admiralty. Since the first weeks of the war, Churchill had taken particular interest in the operations of several armored-car units that had been formed to protect the Naval Air Squadrons based at Dunkirk. As the Western Front settled into the deadlock that would characterize the next four years, he began experimenting with a number of devices to enable the cars to move over rough terrain. But it was not until Churchill saw Hankey's memorandum that he became aware of the possibility of using caterpillar tracks. His imagination was fired by the idea, and on February 20 he established the Landships Committee of the Admiralty to design and develop a machine along the lines proposed by Hankey.[48] For the remainder of the winter and into the spring of 1915, the navy actively investigated various possible designs for the new weapon, and the Royal Commission on Awards to Inventors hardly overrated the importance of this work when it declared that "it was probably due to the receptivity, courage and driving force of Mr. Winston Churchill that the general idea . . . of such an instrument of warfare as the 'Tank' was converted into practical shape."[49]

The army, meanwhile, took no significant interest in the tanks until late June 1915. In February and again in early June, War Office committees did conduct very limited experiments testing the mobility of the Holt tractor, but the trials were judged failures and not followed up. Swinton, totally unaware of the navy's activities, continued to pursue his project with senior army officers whenever possible but with little response. Finally on June 1, he submitted a memorandum to the subchief of staff of the BEF describing in more detail than ever before the specifications and uses of his new weapon.[50] Calling it an "Armoured Machine Gun Destroyer," he specified that it should be armed with two Maxim machine guns and a two-pound cannon, able to cross a four-foot side ditch, climb in and out of a "broad cavity," and go over breastworks. He stressed that the new machines should be employed only when they were in sufficient numbers to be decisive—he suggested an attack employing fifty tanks along a three-mile front—and that they should be assigned, as much as possible, to attack specific enemy machine-gun positions. The initial reaction to this proposal was generally unenthusiastic, but an Inventions Committee formed at BEF headquarters a few days later was more receptive. The committee forwarded the memorandum to the commander of the BEF, Sir John French, who in turn forwarded it to the War Office on June 22, with a covering letter stating in part: "There appears to be considerable tactical value in this proposal."[51]

During the summer of 1915, the pace of development increased rapidly. At the end of June, the Landships Committee was made a joint army-navy committee and in mid-July, Swinton was recalled to London to replace Hankey temporarily while the latter went to the Dardanelles to report on the crisis in the Middle East. For the first time, Swinton learned fully of the work being done by the Admiralty, and in early August he met the men who had been working on the project.[52] On August 28, a meeting was held to coordinate the efforts of the army and navy. There, they agreed that the Admiralty would continue its work based on War Office specifications until they went "sufficiently far"—a term not defined—and the work would then be transferred to the Ministry of Munitions.[53]

On September 19, trials of a prototype vehicle were held for the Landships Committee. The model, named the "Tritton," was unable to meet any of Swinton's specifications, but following the tests, he was shown a wooden mock-up of a new model that seemed to have more promise. This model, named the "Centipede," had a lozenge shape, an upturned nose, and caterpillar tracks that circumvented the hull enabling it to climb a height of five feet and cross a gap of nine. Ten days later, it was decided to push ahead with the new model and drop the Tritton.[54] Thus was born the Mark I tank, known throughout the British Army as "Big Willie."

The year between this decision to develop the Mark I and its debut on the battlefield was marked by uneven and often frustrating progress. The tank introduced innumerable technical problems. The army was unable to provide suitable guns and the Landships Committee ultimately turned to a six-pound naval gun. Steering difficulties led to the addition of two large wheels mounted at the rear of the rank, but these proved so cumbersome that the solution was almost as bad as the problem. Although these wheels were dropped in later models, they were retained in the Mark I, producing many difficulties. To make matters worse, deficiencies in the British metallurgical industry produced iron and steel of such poor quality that from thirty-three to forty percent of all track links were useless.[55] In spite of these numerous difficulties, the first Mark I was ready for testing in February 1916. At first sight, it was as unlikely a candidate to win the war as could be imagined. Large and ungainly (it measured approximately twenty-seven by thirteen by eight feet high and weighed twenty-eight tons), it could attain a maximum speed of only three and a half miles per hour with a twelve-mile radius of action. After traveling seventy miles, the engine had to be completely overhauled. The crew of eight not only had to deal with the tank's numerous mechnical problems, but also with the heat, noise, and fumes of the engine, which was open and lacked an exhaust system.[56]

The initial trial of the Mark I was held at Hatfield Park near London, where a special course had be constructed to simulate the latest Germany defenses in France. On February 2, a delegation of officials from the government, the War Office, and the BEF gathered to see the tank go through its paces. Among those present were Lord Kitchener, Arthur James Balfour, David Lloyd George, General Sir Williams Robertson, and a very nervous Ernest Swinton. While Swinton had lost none of his faith in his brainchild, he was well aware of its continuing mechanical difficulties and feared a breakdown would discredit the entire project. Fortunately, none occurred on this day. The Mark I performed all of its tasks perfectly, including the crossing of a nine-foot ditch. Most of the spectators, if not enthusiastic, were duly impressed and on February 11, the War Office received a request from the BEF for forty tanks. Swinton, believing this figure was totally inadequate, was able to have the order raised to one hundred.[57]

The decision to order a large number of tanks initiated a period of rapid development for the tank forces. Within a few days, the first tank units were formally created and volunteers solicited from other units in the army. To maintain security, the volunteers were not told the exact nature of their new service until six weeks after they arrived at the tank training camp. The unit itself was given the awkward and uninformative designation of the Armoured Car Section of the Motor Machine Gun Service.[58] It was organized into six companies of twenty-five tanks each with a total strength of twenty-eight officers and 255 enlisted men. Much to his surprise, Swinton was appointed commander of the new unit. In April, the army decided that there should be two types of Mark I's: the so-called Male, armed with two six-pound naval guns for use against enemy machine guns and other fortified positions, and the Female, armed with four Vickers machine guns to protect the Males from enemy infantry moving over open ground to attack them. Each tank unit was assigned an equal number of both types so that they could operate tactically in pairs. By the end of April, a training camp, complete with a large mock battlefield, was ready at Elveden in Norfolk and preliminary weapons and mechanical training began. The first tanks did not arrive until mid-June, however, and large-scale training with the machines began at the end of the month. Meanwhile, the name of the unit had been modified to the simpler but equally uninformative title of Heavy Section, Machine Gun Corps.[59]

Unfortunately, this progress was marred by the emergence of a fundamental disagreement between those who had developed the tank and the army high command. The two had come to a fork in the road, and, apparently without realizing the full implications of their contrasting positions, embarked on separate paths until a deep and bitter chasm separated them. The first step was taken by Swinton in a February 1916 memorandum entitled "Notes on the Employment of Tanks."[60] On the surface it was hardly revolutionary. Swinton limited the role of tanks to "the express purpose of assisting attacking infantry by crossing the defenses, breaking through the obstacles, and . . . disposing of the Machine guns."[61] He made no suggestion that tanks could or should replace either infantry or cavalry, nor did he recommend any radical alterations in strategic or tactical doctrines. He did, however, propose important modifications. In the first place, he asserted that the sector for a tank attack "should be carefully chosen to comply with their [the tanks] limitations, i.e., their inability to cross canals, rivers, deep railway cuttings with steep sides, or woods and orchards."[62] Second, he altered the role of artillery: rather than assisting the infantry by a long preparatory bombardment, it should concentrate on counterbattery fire during the attack to protect advancing tanks. The tanks, in turn, would cut the necessary paths through the wire and eliminate enemy machine-gun positions. Finally, Swinton argued that, since novelty and surprise were the tanks' greatest weapons, "these machines *should not be used in driblets* (for instance, as they may be produced), but . . . the fact of their existence should be kept a secret as long as possible until the whole are ready to be launched together with the infantry assault, in one great combined operation."[63]

Although the reaction of senior army officers to the new weapon ranged from support and even mild enthusiasm to open hostility and skepticism, it is fair to say that virtually no one agreed with these recommendations. Not even the most enthusiastic was willing to risk foregoing the long artillery preparation thought necessary to pave the way for infantry or to change the locale of an attack to suit the tank. Furthermore, by the summer of 1916, the army was engaged in the brutal and costly fighting on the Somme and was under mounting pressure to achieve some success in that sector at the earliest possible moment. In such an atmosphere, expediency rather than careful thought dominated, and in the summer of 1916 expediency dictated that tanks be employed on the Somme as soon as they were available. Since they were supposed to be infantry-support weapons, few questioned that they should be able to support infantry on the latter's terms. This was the vital junction at which the army high command and the tank people parted company: the former believed the tank would have at most a limited effect on the operational methods of the other arms, whereas the latter were already beginning to perceive it would necessitate significant changes in those methods. Without fully realizing it, Swinton had made the tank the fulcrum of his plan of attack and, in so doing, reduced the other arms to the position of auxiliaries to the new weapon. Unfortunately, these differences would grow rather than decline as time passed.

The question of whether to employ tanks on the Somme in only limited numbers came to a head at a July 22 conference between Swinton and representatives of the BEF. In addition to raising the points of his February memorandum, Swinton urged delay on the grounds that neither the tanks nor the crews were ready for combat. The vehicles still suffered from numerous mechanical failures that had to be worked out, and the crews had as yet received only the most rudimentary training in maintenance and negotiating obstacles. The BEF representatives refused to be swayed. They argued that it would be foolish to put further effort into tank development if the weapon was not effective and that the only way to determine this was in combat. Not surprisingly, they won the argument. The conference ended with the agreement that the first tanks would be dispatched to France at the beginning of August, with the remainder following at a rate of twelve per week. They would be employed in small tactical groups of six no later than the end of the summer.[64]

The first tanks left England slightly behind schedule on August 14, and within three weeks, two full companies were in France. In early September, they were ordered to be ready to attack on the fifteenth of the month, and immediate efforts were made to intensify training. But this proved impossible. On a visit to the tank training camp, Swinton was chagrinned to find a carnival-like atmosphere surrounding the new weapon. It was as if they were a new toy; instead of training, the crews were compelled to conduct demonstrations for every unit that happened by. Moreover, these demonstrations were often no more than stunts, such as knocking down trees, that had no tactical value and could only damage the machines. Swinton later recalled:

> The official atmosphere was not very helpful. In many quarters there was an air of amused tolerance or contemptuous skepticism towards the New Arm, in some a tendency to place too much reliance on the influence of a few tanks to make up for the recent disappointment of the offensive. There seemed to be little inclination to make allowance for the imperfections of a brand-new weapon or for the difficulties of those who were handling it in strange surroundings after very little practice.[65]

Despite these problems, the Heavy Section, Machine Gun Corps went into action on September 15 as planned. Two companies—a total of forty-eight tanks—were deployed in support of an attack by seven divisions of the British Fourth Army. Their mission was to destroy enemy bunkers and machine-gun positions so that the infantry could pass through. From the beginning, however, the attack was plagued by problems. Many of the tanks broke down from mechanical problems: of the forty-eight employed, only thirty-six reached their starting points, and only twenty-one crossed no-man's-land to reach the German trenches.[66] Furthermore, only a few of these were able to fulfill their missions because they lacked training with infantry. Nevertheless, those tanks which were able to come to grips with the Germans demonstrated the power of the new weapon. A number of enemy strong points were reduced with only very light casualties and panic was produced among some German tropps. In general, however, the results were far short of Swinton's expectations. To the perceptive, there were clear indications of the tank's potential, but these were only a faint glimmer in a generally bleak picture.[67]

The army's reaction to the day's work was generally negative. Rawlinson, commanding general of the Fourth Army, was skeptical at best. He gave the tanks credit for their work against the German strong points, but in his after action report stressed that they had not been sufficiently successful to warrant any changes in standard offensive tactics. Rawlinson's attitude was widely shared and was unofficially confirmed in an October 5 memorandum from BEF headquarters.[68] Haig's reaction, however, was more positive. While he did not question the need to retain established tactical doctrines, he saw a significant role for the tank within them. On the day of the attack, he wrote in his private diary, "Certainly some of the tanks have done marvels and have enabled our attack to progress at a surprisingly fast pace."[69] A few days later, he told Swinton, "Though the tanks had not achieved all that had been hoped, they had saved many lives and had fully justified themselves; that he wanted five times a many; and that I should go home and continue to command, raise and train the force."[70] On September 19, Haig presented the War Office with an order for 1,000 additional tanks and called for an expansion of the Heavy Section. The War Office balked at the size of this request, but Lloyd George intervened and it was approved.[71]

From the fall of 1916 to the spring of 1917, the Heavy Section underwent a radical transformation as new units were raised and new personnel brought in to man them. In October, the four companies in France were expanded to four battalions of three companies each, and the two companies in England were reorganized to form the nucleus for five additional battalions. Beginning in January 1917, the battalions were organized into brigades of two battalions each. The composition of the battalions also underwent a series of changes. Eventually, a structure of three companies each with twelve tanks for operations and four for training was adopted. At the same time, the new Mark IV tank replaced the Mark I. Identical in outward appearance to the earlier model, with the exception of the "tail," the Mark IV was much more reliable and efficient. In addition, the Male version was now equipped with four machine guns plus a cannon and collapsible sponsons on each side for easier movement by rail. Finally, the Heavy Section went through two further changes in name: in November 1916, it was designated the Heavy Branch, Machine Gun Corps and in January 1917, simply the Tank Corps.[72]

This expansion necessitated the influx of new personnel at the command level as well. The commitment of tank units to combat in France obviously necessitated the appointment of a field commander as well as an administrative commander in England, and in October Lieutenant Colonel Hugh Ellis was appointed to direct operations at the front. Ellis lacked prior experience with tanks, but he was recommended by Swinton because he was an excellent officer and well-versed in the political intricacies of BEF headquarters. He proved an excellent choice. A few days later, the corps was shocked when Swinton was transferred from his position as administrative commander in England back to Hankey's staff. The ostensible reason for this move was that Swinton's talents could be put to better use in the new position since the old one had become largely routine, but many of the tank people believed that the real cause was that he had ruffled too many feathers in high places. This apparent prejudice was unfortunate—even though his replacement proved quite capable—because it cost the corps its most experienced and farsighted officer. A final appointment, totally without significance at the time, was the addition of Lieutenant Colonel J. F. C. Fuller to the staff of the corps in December 1916. Although he, too, had never worked with tanks before and was frankly skeptical of their value, he quickly became the brains of the unit, planning attacks, developing tactics, and writing a long series of valuable memoranda on the lessons of its combat experience.[73]

Throughout this period of reorganization, the Tank Corps was largely removed from combat operations, but at the battle of Arras in April 1917, it resumed a major combat role. Unfortunately, the army still refused to change its offensive tactics. In some areas, the tanks were employed where the terrain was unsuitable; in other areas, they were employed in too few numbers to achieve anything more than minor successes. They also continued to be plagued with mechanical problems. In an attack on April 23, only fifty percent of the tanks employed reached starting positions. As on the Somme, there were striking individual successes, but their overall impact was minor. The situation worsened considerably during the Flanders offensive, which opened in June. Although 216 tanks were gathered for the assault—the largest concentration to date—they again were employed on unsuitable terrain and in a role that greatly reduced their chances of success. Ellis voiced "misgivings" over the length of the preparatory bombardment to friends at BEF headquarters and made stronger appeals to withdraw the tanks after the battlefield became a morass of mud in August, but to no avail.[74] The army stubbornly attempted to push the tanks through the battlefield and severely criticized them when they failed. Swinton undoubtedly expressed the bitterness of many in the corps when he later wrote:

> Nothing can be said in defense of the fatuous employment of machines, weighing over thirty tons, in the liquid mud of what degenerated

> into a swamp battle, or of the subsequent and equally fatuous condemnation of them for not functioning well under these conditions.[75]

Fuller, meanwhile, remained at the Tank Corps headquarters concentrating his fertile imagination on an operation more suitable to the new arm. The result was a plan constructed completely around the capabilities and limitations of the tank. To minimize the tanks' difficulties in crossing rough ground, he chose the Cambrai sector east of the old Somme battlefield, where the terrain was open and rolling and had not been chewed up in previous operations. He eliminated traditional preparatory artillery bombardment designed to clear paths through the wire and destroy enemy strong points and gave this role to the tanks. He assigned the artillery the task of covering the roads leading into the battle area and preventing enemy reinforcements from moving over them. He planned the movements of the tanks with the greatest care. The German defenses in the Cambrai sector were part of the Hindenburg Line and consisted of three main lines and one uncompleted support line, all of which were heavily covered by wire and situated on reverse slopes to minimize the effect of Allied artillery. To deal with this system, Fuller divided the attack sector, about six miles wide, into a series of objectives and subdivided these into "tank section attack areas," based on the number of strong points in each area. These areas were assigned a section of one Mark IV Male and two Mark IV Females together with an infantry force. To cross the antitank ditches, which ranged up to twelve feet wide, each tank carried a fascine to be laid at specific points over which the section could then cross. The attack was designed so that successive waves of tank-infantry teams would leapfrog through the German defensive system, overrunning positions before the enemy reserves could make their way through the Allied artillery cordon.[76]

Fuller's plan was originally designed as a raid to inflict casualties and destroy equipment rather than to gain ground, for both he and Ellis realized that the high command would never approve a major operation that could divert British forces from Flanders. The project was well received by the army commander in the Cambrai sector, but his desire to expand the attack in a major assault to improve his position caused Haig to turn it down, as Fuller and Ellis foresaw. When the Flanders offensive ended, however, Fuller's plan was revived and on October 20, it was approved for the following month. Six infantry divisions, nine tank battalions, 1,000 guns, and a cavalry corps were assigned to the assault.[77]

The battle of Cambrai opened at 0620 hours on November 20, 1917.[78] The British achieved almost complete surprise, and by 1600 hours had overrun all three German defensive lines with amazingly small losses except in one sector, where, contrary to their training, the tanks and infantry had become separated. Only the lack of reserves and inability of the cavalry to keep up with the advance prevented the British from achieving their cherished breakthrough. Nonetheless, their success was striking: at a cost of fewer than 4,000 casualties, the British had advanced up to 10,000 yards on a front of 13,000 yards and captured 8,000 prisoners and 100 guns. For the first time since the war began, churchbells in England chimed a message of victory.

Although a surprise German counterattack ten days later recaptured most of the territorial gains of November 20, the lessons of the battle were clear. No less a person that Haig, after noting how the tanks had enabled the British to dispense with the long preparatory bombardment that forfeited surprise, concluded that "the great value of the tanks in the offensive has been conclusively proved."[79] On November 27, a proposal to double the Tank Corps to eighteen battalions, which had been approved the previous June but canceled in July because of manpower shortages, was reapproved. By February 1918, the strength of the corps stood at thirteen battalions organized into five brigades.[80] The power of the corps was further increased by the addition of two new tanks, the Mark V and the Medium A or Whippet. The Mark V was identical in outward appearance to its predecessor but vastly superior in reliability, so much so that at the battle of Hamel in July, all Mark V's were on the jump-off line on time.[81] The Whippet, on the other hand, was designed to exploit the success of the heavier tanks. It was both smaller and lighter than the Mark IV or Mark V, armed only with a light machine gun and capable of crossing only small trenches, but it had a maximum speed of slightly over eight miles per hour and a forty-mile radius of action.[82] Although no one seemed to recognize it at the time, this was the first direct challenge to horsed cavalry.

This expansion was halted suddenly by the opening of the massive German offensive of March 21, 1918. With stunning rapidity, the British lines were shattered and their forces hurled back. Numerous tanks were abandoned for lack of spare parts, and those which could fight proved to be too slow to counter the German advances. On March 26, however, the first Whippets went into combat, and, although their success was limited, their speed was useful in slowing down the Germans until infantry reserves could be brought up. Generally, however, the Tank Corps played an insignificant role in stemming this last German effort, which led many senior British officers to conclude that, while tanks were useful on the offense, their value on defense was minimal. As a result, two tank battalions then being raised in England were stripped of most of their personnel to provide infantry replacements, and the corps in France was ordered to be reduced to four brigades to provide some 2,600 additional replacements. Fortunately, Winston Churchill, newly returned to the government as secretary of state for war, opposed this reduction and had it delayed until the crisis was over. By June, the tide had definitely turned, and the BEF, preparing for the offensive once again, not only canceled its request for the reduction of the corps, but also agreed to its expansion to thirty-four battalions.[83]

The last great British offensive of the war opened in August on a thirteen-mile sector near Amiens, astride the Somme River and only a few miles west of the 1916 battlefield. For this attack Haig concentrated 257,562 men, nine battalions of Mark V's, 600 aircraft, and 2,070 guns in General Rawlinson's Fourth Army. In addition, three infantry divisions, three cavalry divisions, and two battalions of Whippets were in reserve to exploit any success. The total tank strength of these forces was 580, of which 420 were assigned to the initial assault. Opposing them stood the German Second Army, a sadly depleted force of eleven divisions that averaged only 3,000 effectives each.[84] Unlike their previous major offensives, the British placed a great premium on surprise during the preparations for this battle. All troop movements to the front were conducted at night, the assaulting troops were not told of the plans until the last possible moment before the attack, and the artillery was registered alternately so there was no increase in fire to warn of the impending attack.[85] Significantly, the long artillery preparation characteristic of earlier offensives was replaced by a short, intensely fierce bombardment supplemented by reliance on tanks. The battle of Amiens was, in fact, designed to be a large-scale Cambrai.

The first day of the British offensive, August 8, was justly described by General Ludendorff as "the black day of the German army."[86] The British surprise was complete and by dusk they had penetrated almost to the third German line eight miles from their jump-off positions, inflicted some 27,000 casualties on the enemy, and captured 400 guns and huge quantities of machine guns, trench mortars, and other equipment. British casualties were extremely light, but on the following day, only 145 tanks were fit for action.[87] For three more days, the British continued to drive the Germans back, though with steadily increasing losses and decreasing results. On August 11, despite strong French opposition, Haig halted the attack and began shifting his forces for a fresh blow elsewhere. His losses for the four days totaled about 27,000 men, scarcely the German casualties for the first day alone; but of the 688 tanks sent into action, 480 were turned over to salvage and only about thirty were fit for immediate action.[88] Perhaps even more significant than this success was the change the battle of Amiens produced in the army's attitude toward tanks. General Rawlinson, long skeptical of their value, best reflected this change in his Special Order of August 16, in which he declared: "The success of the operations of August 8 and succeeding days was largely due to the conspicuous part played by the 3rd, 4th, and 5th Brigades of the Tank Corps."[89]

During the last three months of the war, the British assaulted up and down the Western Front, following the pattern established at Amiens. On each occasion, the artillery provided a short but intensive bombardment, massed tanks led the infantry through the German network of trenches and barbed wire, and the attack was halted when its momentum slowed and the maximum enemy casualties had been inflicted at the lowest cost. Although the degree of success varied from attack to attack, it was readily apparent to all that the years of deadlock were over. On August 21, the British opened the battle of Bapaume with 190 tanks. This time, the Germans placed their artillery well to the rear and concentrated on destroying British tanks once they had penetrated the German defensive zone. Although this increased tank losses, it failed to stop the advance.[90] On September 27, the British launched a direct attack on the Hindenburg Line led by 230 tanks. Within two weeks, they had advanced up to twenty miles on a thirty-mile front and taken 48,000 prisoners and 630 guns. Open country now lay ahead, but the infantry and the Tank Corps were exhausted and the cavalry, the only fresh troops available, were still too vulnerable to German machine guns.[91]

Although the fighting continued for three more weeks, the assault on the Hindenburg Line was the last major British offensive of the war. The Germans continued to withdraw to shorten their line and without a period of respite, the British could do little more than hasten them. By early November, the Tank Corps was so reduced that sections and companies were doing work previously assigned to companies and battalions. Half of the 1,993 tanks and armored cars employed since August had been sent to salvage and, while only a few were totally scrapped, only 204 were returned to duty by the time of the Armistice.[92]

Lessons of the World War: The Official View

The British Army's interpretation of the lessons of World War I may be stated quite simply: it was that the ultimate Allied victory, however costly, had vindicated completely the strategy of attrition and that this doctrine should be strictly maintained in the army's postwar training and planning for future war. Field Marshal Haig set the tenor for the army's position in this final dispatch summarizing the war as whole:

> Neither the course of the war itself nor the military lessons to be drawn therefrom can be properly comprehended, unless the long succession of battles commenced on the Somme in 1916 and ended in November of last year on the Sambre are viewed as forming part of one great and continuous engagement.
>
> To direct attention to any single phase of that stupendous and incessant struggle and seek in it the explanation of our success, to the exclusion or neglect of other phases possibly less striking in their immediate or obvious consequences, is in my opinion to risk the formation of unsound doctrines regarding the character and requirements of modern war.
>
> If the operations of the past 4½ years are regarded as a single continu-

ous campaign there can be recognized in them the same general features and the same necessary stages which between forces of approximately equal strength have marked all the conclusive battles in history. There is in the first instance the preliminary stage of the campaign in which the opposing forces seek to deploy and manoeuvre for position. . . .

Battle having been joined, there follows the period of real struggle in which the main forces of the two belligerent Armies are pitted against each other in close and costly combat. Each commander seeks to wear down the power of resistance of his opponent and to pin him to his position, while preserving or accumulating in his own hands a powerful reserve force with which he can maneuver, and, when signs of the enemy becoming morally and physically weakened are observed, deliver the decisive attack. . . .

In the stage of the wearing-out struggle losses will necessarily be heavy on both sides, for in it the price of victory is paid. If the opposing forces are approximately equal in numbers, in courage, in moral [sic] and in equipment, there is no way of avoiding payment of the price or of eliminating this phase of the struggle.

In former battles this stage of the conflict . . . has often been completed in a few hours. When Armies of millions are engaged, with the resources of great Empires behind them, it will inevitably be long. . . . To this stage belong the great engagements of 1916 and 1917 which wore down the strength of the German Armies.

Finally, . . . the time will come when the other side will begin to weaken and the climax of the battle is reached. Then the commander of the weaker side must choose whether he will break off the engagement, if he can, while there is yet time, or stake on a supreme effort what reserves remain to him. The launching and destruction of Napoleon's last reserves at Waterloo was a matter of minutes. In this World War the great sortie of the beleaguered German Armies, commenced on the 21st March 1918, lasted for four months, yet it represents a corresponding stage in a single colossal battle.

The breaking down of such a supreme effort will be the signal for the commander of the successful side to develop his greatest strength, and seek to turn to immediate account the loss in material and moral which their failure must inevitably produce among his opponents troops. . . . If the whole operations of the present war are regarded in correct perspective, the victories of the summer and autumn of 1918 will be seen to be as directly dependent upon the two years of stubborn fighting that preceded them.[93]

This passage requires little comment; it is—and no doubt was intended to be—a frank, unapologetic reaffirmation of faith in the Napoleonic model of war. The only difference between earlier battles and that on the Western Front was the length of the latter and this, Haig explained, was because of factors over which the army had no control: England's state of unpreparedness in 1914; setbacks in other theaters, particularly Russia and Italy; the problems of coordinating the activities of different nations; and the size of the armies and the power of modern defenses. Given these problems and the situation facing the Allies, he concluded, "Our total losses in the war have been no larger than were to be expected."[94] Thus was the nature of modern war.

Haig's sentiments were echoed in postwar British military journals with increasing stridency as criticism of the army's conduct of the war grew. In 1921, an *Army Quarterly* editorial declared that the army's critics failed "to grasp that the true results of the earlier [1914–17] Allied offensives are not to be measured by the amount of ground gained, but in the exhaustion of the enemy."[95] It returned to the subject in 1925, stating that the purpose of those offensives had been to destroy the enemy's will to resist and for this: "The surest way of destroying the will to resist of commanders and their staffs has hitherto been to confront them with their inability to find reserves to parry fresh blows."[96] And in 1928: "We may deplore the losses, but of the effect on the enemy there is now [in 1928] no doubt. The Somme broke the spirit and morale of the German Army and destroyed the old trained cadres on which it relied."[97] One officer (who, it should be noted, wrote a number of articles favoring mechanization) undoubtedly expressed the feelings of the overwhelming majority of the officer corps when he declared: "Indeed, if we ever have the misfortune to take part again in a great war we shall have grounds for congratulations if we find ourselves as well prepared in point of doctrine and training as we were in 1914."[98]

The greatest support for this interpretation of the war, however, appeared in the *Official History of the Great War,* published under the auspices of the War Office. Comprising fifty massive volumes containing an extraordinary amount of detail on every British campaign in the war, this series was and continues to be the subject of considerable controversy. Critics claim that it contains numerous distortions, inaccuracies, and downright fabrications; supporters say that it is a work of the highest scholarship and unsurpassed for studying British operations in the war. The truth does not, as is so often the case, lie somewhere in the middle; rather, it lies in how one views the war, for these volumes are not simply an account of that conflict, but an account from the perspective of British military doctrine. The history certainly did not ignore mistakes by the army's leadership. In its account of the battle of the Somme in 1916, for example, it noted that in the beginning senior commanders were inexperienced in handling large bodies of troops and overrated the effectiveness of their artillery preparation; that in the attack of September 15, in which tanks were used for the first time, the coordination of infantry and tanks had received too little attention and the decision to employ tanks all along the front was poor because it prevented tank commanders from effectively controlling their units; that commanders often ordered infantry to take positions "at all costs," producing unnecessarily heavy casualties; and that corps commanders positioned their headquarters only with reference to other headquarters and were thus not in a position to effectively control their troops engaged at the front.[99] Severe as these criticisms may be, they were, nonetheless, confined to matters of detail. On matters of doctrine, the *Official History* was as clear and adamant as was Haig in his final dispatch—and just as much in agreement with him.

The foundation of the *Official History*'s interpretation was that "in war the issue between two antagonists who are more or less evenly matched must ultimately be decided by battle," and while it conceded "sound strategy may often limit—but hardly abolish altogether—the necessity of 'wearing down' an opponent by sheer hard fighting," it concluded that the continuous front and "character and temper of the German resistance made inevitable battles of attrition."[100] This point was repeated with an almost strident persistance in the description of the Somme offensive of 1916, in which British losses reached their peak, and was used to justify the claim that "the Germans of 1916 were such doughty opponents that nothing but continuous hard pounding could bring them to this state [of moral collapse]."[101] In this light, the *Official History* concluded that "it is not too much to claim that the foundations of the final victory on the Western Front were laid by the Somme offensive of 1916."[102] To substantiate this, the *Official History* presented a wide array of facts and figures. It pointed to the German peace proposals of December 1916 and the restoration of unrestricted submarine warfare as signs of German desperation. The German withdrawal to the Hindenburg Line was interpreted, not as a carefully conceived strategic measure, but as a desperate withdrawal forced on Hindenburg and Ludendorff by severe manpower shortages. It specificially referred to no fewer than seven German unit histories and memoirs expressing the belief that, in the words of one, "the Somme was the muddy grave of the German field army, and of the faith of the infallibility of the German leaders, dug by British industry and its shells." Its most controversial contention, however, was that the battle of attrition favored the Allies more than the Germans and that "after 1916, the year of Verdun and the Somme, the military value of the German officers and men had fallen faster than that of the French and British, whose natural fighting qualities now turned the scale."[103] In the eyes of the *Official History,* the greatest tragedy in this chain of events was not the human cost of these successes, but the fact that the generals were not permitted to take advantage of this situation in 1917:

> The concealment by the Germans of their true losses and the admittedly heavy French and British losses, combined with the attitudes towards their generals of the Allied governments which, like others before them, wanted victories without bloodshed—with the result that they get bloodshed without victories—led to the moral and material advantage gained over the enemy in 1916 being wasted in 1917.[104]

This is not to say that the army ignored the importance of tanks within the strategy of attrition. As the evaluations of Rawlinson and Haig on the contributions of tanks in the final offensives of the war demonstrate, the very opposite was true. By cutting paths through enemy barbed wire and eliminating enemy strong points, tanks had enabled infantry to attack without the long artillery preparation that had meant the loss of surprise and to advance at less cost to themselves while inflicting greater casualties upon the enemy. Thus, in the army's view, tanks reinforced rather than undermined the validity of the strategy of attrition because it made such a strategy more efficient. This was the contention that the theorists of armored war challenged and which forms the focus of the tank controversy in the British Army during the interwar years.

Notes

1. The best introductions to Clausewitz are Hans Rothfels, "Clausewitz," in *The Makers of Modern Strategy,* ed. Edward Mead Earle (New York: Antheneum, 1967), pp. 93–113; and Roger A. Leonard, ed., *A Short Guide to Clausewitz on War* (New York: Capricorn Books, 1968), pp. 3–37. On Clausewitz's later influence, see Peter Paret, "Clausewitz and the Nineteenth Century," in *The Theory and Practice of War,* ed. Michael Howard (London: Cassel and Co., Ltd., 1965), pp. 23–41.
2. Michael Howard, "Strategy and Policy in the Twentieth Century Warfare," *Studies in War and Peace* (London: Temple Smith, 1970), p. 187.
3. Carl von Clausewitz, *On War,* trans., J. J. Graham with introduction and notes by F. N. Maude, 3 vols. (London: Kegan, Paul, Trench, Trahner and Co., Ltd., 1918), 1:44–45. Although this translation is awkward, often difficult to read, and contains many errors, it is used here because it was the only version available in English before 1914. A more recent and far superior edition is now available, translated and edited by Michael Howard and Peter Paret (Princeton, N.J.: Princeton University Press, 1976). All references here are to the Graham and Maude edition.
4. Ibid., p. 270.
5. Ibid., p. 285.
6. Ibid.
7. Ibid., p. 192; 2:5. Italics in original.
8. Ibid., 1:84.
9. Ibid., p. 246.
10. Ibid., pp. 286–88.
11. Jay Luvaas, "European Military Thought and Doctrine, 1870–1914," in *The Theory and Practice of War,* ed. Michael Howard (New York: Frederick A. Praeger, 1966), p. 71.
12. Michael Howard, "Armed Forces and the Art of War," in The New Cambridge Modern History, vol. 11: *Material Progress and World-Wide Problems, 1870–98,* ed. F. H. Hinsley (Cambridge: Cambridge University Press, 1967), p. 217.
13. Theodore Ropp, *War in the Modern World* (Durham, N.C.: Duke University Press, 1959), p. 182.
14. Luvaas, "European Military Thought," pp. 84–86.
15. See the comments of Brigadier Dugan following Liddell Hart, "The 'Man in the Dark' Theory of Infantry Tactics and the 'Expanding Torrent' System of Attack," *Journal of the Royal United Service Institution* 66 (February 1921): 22.
16. S. T. Possony and E. Mantoux, "Du Picq and Foch: The French School," *The Makers of Modern Strategy,* ed. Edward Mead Earle (New York: Antheneum, 1967), p. 218.
17. Ferdinand Foch, *The Principles of War,* trans. J. de Morinni (New York: The H. K. Fly Co., 1918), pp. 26–27.
18. Ibid., p. 28.
19. Ibid., pp. 41–42.
20. Possony and Mantoux, "Du Picq and Foch," p. 210.
21. Foch. *Principles of War,* pp. 314–15.

22. Ibid., p. 33.
23. Quoted in Ropp, *War in the Modern World,* p. 204.
24. Alfred Vagts, *A History of Militarism* (New York: Free Press, 1967), pp. 221–22.
25. B. H. Liddell Hart, *The Ghost of Napoleon* (New Haven, Conn.: Yale University Press, 1935), p. 136.
26. Colonel H. Rowan Robinson, "Surprise, Rapidity, Concentration," *Journal of the Royal Artillery* 47 (January 1921): 474.
27. Jay Luvaas, *The Education of an Army* (Chicago: University of Chicago Press, 1964), pp. 179, 197.
28. G. F. R. Henderson, *The Science of War* (London: Longmans, Green and Co., 1906), pp. 68–69, 340–41.
29. Richard Burton, Lord Haldane had already established an outstanding reputation as a legal and educational reformer when he was appointed secretary of state for war in the Liberal government that came to power in 1906. His work in the War Office entitles him, more than any other, to the title "Father of the Modern British Army."
30. Brian Bond, *The Victorian Army and the Staff College* (London: Eyre Methuen, 1972), pp. 259–62. Bond Asserts Wilson did not fully appreciate the effect of modern firepower on tactics, but claims he had too much respect for the Germany Army to believe it could be easily defeated. Further, Bond argues Wilson was more interested in strategy than tactics and largely ignored the latter. For an earlier view of Wilson's role, see Liddell Hart, *The Ghost of Napoleon,* p. 138.
31. Great Britain, War Office, *Field Service Regulations,* Part 1, *Operations* (London: HMSO, 1909). The regulations were amended in 1914, but no significant changes were made.
32. Ibid., p. 128. Italics in original.
33. Ibid., p. 128. Italics in original.
34. Ibid., p. 126.
35. Ibid., p. 127.
36. Ibid., p. 126. Italics in original. It is difficult to weigh exactly the relative importance of the many principles set forth in the *Regulations* that are designed to cover every conceivable type of military operation. The author is aware that it is all too easy simply to extract passages that fit in with the Continental doctrines and has taken this into account in this description. It is significant, however, that many of these passages appear in boldface type in the original text and they are almost the only passages so emphasized. It seems inescapable that they, like Clausewitz's concept of "absolute war," would be the ones to make the greatest impression.
37. The bibliography on this subject could fill a book in itself. Two of the more interesting contributions to the controversy are Leon Wolff, *In Flanders Fields* (New York: Viking Press, 1958), against the generals, and John Terraine, *Ordeal of Victory* (New York: J. B. Lippincott Co., 1963), on their behalf.
38. Quoted from "The Present Situation and Future Plans," a paper prepared by Haig for the War Cabinet in May 1917 and summarized in his diary. Terraine, *Ordeal of Victory,* p. 310.
39. J. F. C. Fuller, *A Military History of the Western World,* 3 vols. (New York: Funk and Wagnalls, 1956), 3: 275; Ropp, *War in the Modern World,* p. 231.
40. Captain Wilfred Miles, *Military Operations France and Belgium: 1916, July 2 to the End of the Somme Battles,* in *History of the Great War Based on Official Documents,* Brigadier General Sir James E. Edmonds, gen. ed. (London: MacMillan and Co., Ltd., 1938), p. 572. The author notes that as late as this battle, some British units still took pride in maintaining the proper alignment while advancing in the assault.
41. Brian Bond, "The First World War," in The New Cambridge Modern History, vol. 12, *The Shifting Balance of World Forces,* ed. C. L. Mowat (Cambridge: Cambridge University Press, 1968), p. 198.
42. Ernest D. Swinton, *Eyewitness* (Garden City, N.Y.: Doubleday, Doran and Co., 1933), pp. 41–43.
43. B. H. Liddell Hart, *The Tanks,* 2 vols. (New York: Frederick A. Praeger, 1959), 1: 10–16. The Wells story appeared in the December 1903 issue of *Strand Magazine.* See also Miles, *Military Operations: 1916, July 2 to the End of the Somme Battles,* pp. 245–46.
44. Swinton, *Eyewitness,* pp. 56–57.
45. Ibid., pp. 60–61. Lord Hankey, *The Supreme Command,* 2 vols. (London: George Allen and Unwin Ltd., 1961), 1: 227–28. A good general account is also found in Liddell Hart, *The Tanks,* 1: 18–43.

46. Hankey, *The Supreme Command,* 1:246.
47. Those members of the Cabinet concerned with the war effort.
48. Hankey, *The Supreme Command,* 1:250 and 2:496; and Swinton, *Eyewitness,* pp. 79–81.
49. Miles, *Military Operations: 1916, July 2 to the End of the Somme Battles,* p. 247, contains the text of the report.
50. Swinton, *Eyewitness,* pp. 106–11, contains the text of the memorandum.
51. Ibid., pp. 125–26.
52. Ibid., pp. 135–38. Swinton first learned of the navy's work in June. The navy personnel were almost in as much dark: although they were working on Swinton's specifications, they had no idea where they came from.
53. Ibid., p. 145.
54. Ibid., pp. 146–50.
55. Correlli Barnett, *The Collapse of British Power* (New York: William Morrow and Co., Inc., 1972), p. 84.
56. J. F. C. Fuller, *Tanks in the Great War* (London: John Murray, 1920), pp. 35–37.
57. Swinton, *Eyewitness,* pp. 164–72.
58. It was originally designated the Tank Detachment, but this was changed within a few days for security reasons.
59. Swinton, *Eyewitness,* pp. 178–201; Liddell Hart, *The Tanks,* 1:44–63; Fuller, *Tanks,* pp. 159–60; Miles, *Military Operations: 1916, July 2 to the End of the Somme Battles,* pp. 245–49.
60. Swinton, *Eyewitness,* Appendix 2, pp. 308–325, contains the complete text of the memorandum.
61. Ibid., p. 308.
62. Ibid., p. 311.
63. Ibid., p. 318.
64. Ibid., pp. 216–17, 222–31. Writing twenty years after the war, Swinton still believed this was a bad decision. Ibid., 248–52. Liddell Hart has agreed with this opinion, both men asserting essentially that it was an early example of the army's misuse of tanks because it refused to accept the advice of the tank experts. The army maintained, however, that the lessons learned from the early employment of tanks were invaluable and it has been supported in this by none other than Fuller. Miles, *Military Operations: 1916, July 2 to the End of the Somme Battles,* pp. 365–66, and Fuller, *Tanks,* pp. 58–59.
65. Swinton, *Eyewitness,* p. 232.
66. Liddell Hart, *The Tanks,* 1:69–74; Fuller, *Tanks,* p. 54, writes that only thirty-two reached the starting point.
67. Fuller, *Tanks,* pp. 54–56; Liddell Hart, *The Tanks,* 1:71–79; Miles, *Military Operations: 1916, July 2 to the End of the Somme Battles,* pp. 363–64.
68. Miles, *Military Operations: 1916, July 2 to the End of the Somme Battles,* p. 367.
69. Douglas Haig, *The Private Papers of Douglas Haig, 1914–1919,* ed. Robert Blake (London: Eyre and Spottiswoods, 1952), p. 167.
70. Swinton, *Eyewitness,* pp. 239–41.
71. Liddell Hart, *The Tanks,* 1:70; Fuller, *Tanks,* p. 61.
72. Liddell Hart, *The Tanks,* 1:54, 82, 90; Fuller, *Tanks,* p. 62.
73. Liddell Hart, *The Tanks,* 1:91.
74. Ibid., pp. 74–104; Miles, *Military Operations: 1916, July 2 to the End of the Somme Battles,* p. 368.
75. Swinton, *Eyewitness,* p. 263.
76. Fuller, *Tanks,* pp. 136–53.
77. Ibid., p. 144. Of these, ninety-eight were for command and supply, leaving 378 for actual combat.
78. Ibid., pp. 148–53; Liddell Hart, *The Tanks,* 1:136–48.
79. Douglas Haig, *Sir Douglas Haig's Dispatches,* ed. J. H. Boraston (London: E. D. Dulton, 1920), p. 173.
80. Fuller, *Tanks,* pp. 172–73. It should be noted that not all units were up to full strength.
81. Ibid., pp. 202–3.
82. Ibid., pp. 45–47; Liddell Hart, *The Tanks,* 1:153–56.
83. Fuller, *Tanks,* pp. 64–65; Liddell Hart, *The Tanks,* 1:168.
84. Brigadier General Sir James E. Edmonds, *Military Operations in France and Belgium: 1918, August 8–September 26,* in *History of the Great War Based on Official Documents,* Brigadier

General Sir James E. Edmonds, gen. ed. (London: MacMillan and Co., Ltd., 1947), p. 22; Fuller, *Tanks,* pp. 218–19; Liddel Hart, *The Tanks,* 1 : 174–178.

85. Liddell Hart, *The Tanks,* 1 : 178.

86. Bond, "The First World War," pp. 203–4.

87. Edmonds, *Military Operations: 1918, August 8–September 26,* p. 87; Liddell Hart, *The Tanks,* 1 : 184–85; Fuller, *Tanks,* pp. 224–27.

88. Fuller, *Tanks,* p. 227.

89. Ibid.

90. Ibid., pp. 250–59.

91. Ibid., pp. 275–76; Liddell Hart, *The Tanks,* 1 : 189–92.

92. Fuller, *Tanks,* pp. 286–87.

93. Haig, *Dispatches,* pp. 319–21. This was contained in his dispatch 8, March 21, 1919, surveying the course of the campaign in France.

94. Ibid., pp. 321–23.

95. Anonymous, "Editorial," *Army Quarterly* 2 (July 1921): 225.

96. Anonymous, "Editorial," *Army Quarterly* 11 (October 1925): 185.

97. Anonymous, "Editorial," *Army Quarterly* 16 (April 1928): 6.

98. Lieutenant Colonel H. Rowan Robinson, "Surprise, Rapidity, Concentration," *Journal of the Royal Artillery* 47 (January 1921): 474.

99. Brigadier General Sir James E. Edmonds, *Military Operations France and Belgium: 1916 to July 1, The Somme,* in *History of the Great War Based on Official Documents,* Brigadier General Sir James E. Edmonds, gen. ed. (London: Macmillan and Co., Ltd., 1932), pp. 482–84; Miles, *Military Operations: 1916, July 2 to the End of the Somme Battles,* pp. 314–315, 563–64, 569.

100. Miles, *Military Operations: 1916, July 2 to the End of the Somme Battles,* p. 561.

101. Edmonds, *Military Operations: 1916, to July 1, the Somme,* pp. vii–viii, *484.*

102. Miles, *Military Operations: 1916, July 2 to the End of the Somme Battles,* p. xiii. This preface was written by Edmonds.

103. Captain Cyril Falls, *Military Operations France and Belgium: 1917, The German Retreat to the Hindenburg Line and the Arras Battles,* in *History of the Great War Based on Official Documents,* Brigadier General Sir James E. Edmonds, gen. ed. (London: Macmillan and Co., Ltd., 1940), pp. vii–ix; Edmonds, *Military Operations: 1918, August 8–September 26,* p. 15. The German version of the effect of the Somme battles is quoted in Edmonds, *Military Operations: 1916, to July 1 the Somme,* pp. 494–95.

104. Falls, *Military Operations: 1917, The German Retreat to the Hindenburg Line and The Arras Battles,* IX.

• 3 •

The Challenge: The Theory of Armored Warfare

Introduction

WORLD War I divided the people of Europe into two diametrically opposed camps: those who believed that the war and its accompanying problems were the result of the incomplete application of nineteenth-century values and principles and those who believed that they were a product of the inherent fallacies of those values and principles. To the former, the postwar era provided an occasion for further reform to complete the work in process before 1914. They saw hopeful signs in the creation of new nation states in Eastern Europe, the spread of democratic governments, and the establishment of the League of Nations. Their opponents, on the other hand, regarded this time as an opportunity to complete the destruction of nineteenth-century civilization, which they held responsible for the war and its misfortunes. Such people turned to fascism, communism, or a host of lesser and short-lived radical movements to achieve their ends. Within the British Army, a parallel phenomenon occurred. On the one hand were those who continued to adhere to the strategy of attrition and maintain that many of the problems of the war had arisen because the generals had not been permitted to apply it without interference. They admitted the generals had erred badly in judging the time required to defeat the Germans, but beyond this they would not go; modern war, as Clausewitz had written, was a bloody business and the only way to limit the bloodshed was to destroy the enemy forces as soon as possible. Opposed to them were those who argued that the prewar doctrines were inherently false and that a new concept of military thought and action had to be constructed lest what remained of European civilization be destroyed in another war. It was from this latter assumption that the theory of armored warfare emerged in England.

The Tank Advocates

The men most responsible for the development of this theory were Major General J. F. C. Fuller and Captain Sir Basil H. Liddell Hart. Together with a small group of adherents, they developed the concept, worked out the technical details of its execution, and propagandized their ideas to both the army and the public at large. Before examining their ideas, however, it is necessary to know these men in order to establish what distinguished them from so many of their contemporaries and to understand the role their personalities played in the British Army's reception to their ideas.

John Frederick Charles Fuller was born on September 1, 1878,[1] and according to his own, perhaps fanciful, recollection, was a skeptic by the age of five. He developed an early interest in the occult, and his teenage letters to his parents were bordered with supernatural creatures and symbols that, although revealing some artistic ability, cannot but lead to speculation as to the workings of his mind. Perhaps they are simply an early indication of his lifelong pleasure at shocking people, but whatever the explanation, it is not every son who sketches luried birds of prey with bloodied talons on letters to his mother![2] Fuller entered the British Army at the age of twenty, following a year at Sandhurst, and his early career followed the pattern of the period. Tours in India alternated with assignments at home, and except for service in the Boer War, he saw no significant combat. Off duty, however, he set himself apart from his fellow officers. He never seemed to have appreciated their life-style, and, instead of pursuing the finer activities of a gentleman, he used his free time to study philosophy and, of all things, the art of war. His tours of India also aroused an interest in Oriental religions that eventually led him to write two books on the subject.[3] It is probable that, while these early studies provided the foundation for his later writings, they also produced an estrangement from his fellow officers and a profound sensitivity in his relations with them. He rationalized these differences by regarding himself as their intellectual superior and, while there is much truth in this, it led to an unfortunate arrogance on his part, a tendency to view anyone who did not see things exactly as he as a complete idiot.

When World War I broke out, Fuller was a student at the Staff College and, with the rest of his class, was sent to the points of embarkation to assist loading the BEF for France. According to one of his classmates, Fuller spent the time until the troops arrived poking about the station, where he discovered that the railway tracks at the embarkation point were so constructed that the trains would not be able to move in and out smoothly: they could enter the station with no problem, but additional tracks were needed so they could leave without creating a bottleneck. Fuller pointed this out to the railroad authorities, who corrected the matter, and, while the effect of these changes is not certain, his classmates felt a sense of pride that one of their number had helped get the Expeditionary Force to France on schedule.[4] Fuller did not reach the Western Front until 1915, and for the next eighteen months he served in various staff positions with infantry units. In December 1916, in what was certainly the turning point of his life, he was posted to the staff of the Heavy Branch, Machine Gun Corps. He served with corps until mid-1918 and was eventually promoted to GSO 1, or chief of staff, where he planned the Cambrai attack of November 1917. The following May he devised an operation to win the war in 1919 with a massed tank attack aimed at the German lines of communications and headquarters units.

Fuller's memoirs provide a fascinating insight into his own thoughts during these years. He was not much of a combat officer; rather, he was what today would be called an "idea man." His memoirs are almost wholly a summary of his papers and staff studies devoted to the employment of tanks and other tactical and strategic problems facing the army. They have about them an air of certainty so characteristic of him that it is easy to picture an officer at higher headquarters receiving one of these papers and sighing wearily, "Another one of Fuller's plans to win the war in the next week!"

In July 1918, Fuller was assigned to the War Office—which he derisively call the Tower of Babel—as deputy director of staff duties (tanks) and in 1920, returned to the Staff College as an instructor. During these years, he firmly established his reputation as the leading advocate of the new arm, the *Army Quarterly* saying of him in 1921 that he "requires no introduction as an exponent of his subject."[5] Although he was often chided for overstating his case, he was publicly treated with considerable respect, and his books and articles were recommended for officers who wished to keep up with their profession. By the mid-1920s, however, Fuller had come to believe that his ideas were ruining his career, as he had dared to disagree with the army establishment. In 1924, he wrote Liddell Hart that he expected to be shuttled between bad posts and half-pay until he reached the rank of major general, at which time he would be forced on half-pay until he retired.[6] Except for one brief interval, this forecast proved accurate: when Sir George Milne was appointed chief of the imperial general staff in 1926, he selected Fuller as his special advisor on mechanization, and in 1927 he appointed him to command the Experimental Mechanized Force, which was then being formed. The details of the controversy that erupted in the wake of this appointment will be discussed later[7]; all that need be noted here is that Fuller eventually refused the post and that this marked the end of his career. As he predicted, he went on half-pay in 1930 and retired three years later.

In retirement, Fuller not only took up writing full time, but also became involved in political activities that did his reputation irretrievable damage and his cause little good. Fuller had always been authoritarian by disposition, but the British government's failure to provide the dynamic leadership he felt necessary for the country's survival (and, of course, to carry out the military reforms he advocated), coupled with the apparent dynamism of Germany and Italy, now led him to become a fascist. The transition was completed after his retirement from the army in 1933, when he openly joined Sir Oswald Mosley's British Union of Fascists and became one of their most prominent spokesmen. The effect of this transformation on Fuller's military writings was twofold: first, it led him to emphasize the Social Darwinist view that war was a positive force in the evolution of society; and, second, it led him to conclude that democracy was responsible for the development of total war. With regard to the first, he maintained that war forced a society to live off its essentials: the strain of war forced the participants to look at themselves and discard those ideas and traditions which were no longer valid. "War is a great physician," he had written in 1923, "a great medicine, a great purge."[8] By the 1930's, he was arguing that it brought out the best in society, those noble acts which raised a people to great heights and enabled them to survive.[9] Fuller's belief in the relationship between democracy and total war reflected his sense of elitism and authoritarianism. Democracy, he reasoned, meant government by the most ignorant; consequently, when a democracy went to war, passions of the mob quickly took control of policy and the war became an emotional rather than a rational act. This, he concluded, had been the course of history since the French Revolution and, unless changed, would lead to the destruction of European civilization.[10]

Even from this cursory outline, it is clear that Fuller's army record gives him no claim to fame. The promise of greatness was there—his work with the Tank Corps demonstrated that—but it was a potential never fully realized. Fuller's reputation, then, must rest on his writings, a fact with which he most certainly would have agreed, as his memoirs contain far more about the development of his theories of war than they do about his activities. He was blessed with an extremely active and probing mind and took to the pen more readily that most of his fellow officers took to the sword. His first book, *Training Soldiers for War,* was published in 1914 and after the war he produced a remarkable steady stream of books and articles. In 1919, he won the Military Gold Medal for the outstanding essay presented in competition to the *Journal of the Royal United Service Institution* and the following year won a similar prize in the naval competition.[11] His range of interests was incredibly broad even within the field of military literature, and he produced numerous works still considered important in the fields of military theory, history, and the problems of imperial defense. It is doubtful if he ever thought that there was any problem that, given sufficient time to study, he could not solve. Despite his great activity and often penetrating insight, however, he failed to win many converts within the army.

Unquestionably, much of the answer for this failure lies in Fuller's personality. In 1923, he admitted to Liddell Hart:

> I have never attempted to make an appeal to the people in order to gain a following. It amuses me to state what I believe to be true, but whether my audience understands me or not I do not much care because the truth in the end wins through.[12]

This rather cavalier attitude permeated much of Fuller's writing, but what proved even more damaging is the fact that he often went beyond blunt honesty and took great relish in deliberately antagonizing his readers. Examples of this intent abound throughout his writings, but few match the biting sarcasm in his description of a series of exercises held shortly after the end of the war. In the early 1920s, the army developed and ordered production of an antitank gun and decided that, until there was a sufficient number, available flags would be used in their stead during field training. Such a technique has often proven valuable if properly applied, for it provides both commanders and troops with experience in the tactical employment of the weapon. In this instance, however, those directing the training too often lost their sense of reality; whenever a tank came within the maximum effective range of the theoretical weapons, it was automatically destroyed. The absurdity of this practice is obvious, but Fuller's criticism could not have been written in a way more calculated to arouse hostility:

> In 1918, the Germans, possessing thousands of guns and hundreds of anti-tank rifles, did not stop our tank attacks; foolishly, it seems, they never tried flags; yet, had they done so, according to the theory of our Higher Command, there can be little doubt that today the imperial standard of the Hohenzollern would be floating over Buckingham Palace.[13]

The reaction of the War Office to such sarcasm can easily be imagined, but expressions of dismay at Fuller's literary excesses were not confined to these quarters. On one occasion, none other than Ernest Swinton called Fuller "d——d silly" and noted that he had gained a sort of buffoon reputation; indeed, Liddell Hart's correspondence contains a number of sharp criticisms of Fuller by other tank advocates.[14] Needless to say, those opposed to the new theory of armored warfare found Fuller completely unbearable. In 1926, General Sir Archibald Montgomery-Massingberd, answered a query from Liddel Hart as to whether he had read Fuller's *The Foundations of the Science of War,* which had just been published, with a resounding "No. I have not read Fuller's book! And I don't expect I ever shall. It would only annoy me."[15] This antagonism was based less on a reaction to Fuller's message than on his style of writing, a fact the general made clear in a later letter in which he complained that "Fuller's mode of writing and assumption of infallibility and mental superiority renders it impossible to take him seriously or derive any pleasure from reading what he writes." He later summed up his opinion of Fuller's books by saying that "they are written with the intention of annoying someone."[16] Although this was a most regrettable reaction, Fuller's attitude reveals that, under the circumstances, it was neither unfounded nor unfair. Even Liddell Hart, while defending his close friend, had to concede that some of these accusations were valid, pleading only that readers try and forget Fuller's manner and concentrate on his ideas. In responding to Montgomery-Massingberd's letters, for example, he admitted that, while Fuller could be "astonishingly brilliant" in places, he could also be "obscure or even wearisome elsewhere." As for Fuller's personality, Liddell Hart again confessed: "I used to find him very irritating," but added that "as I came to know him intimately I got below his manner to the real man. The kernel of the matter is that he is in the direct line of the race of reformers all of whom were intolerant and most difficult to live with."[17]

It is obvious that the subject of Fuller's personality is an important element in understanding at least some of the opposition the tank advocates faced. As the group's most prominent spokesman, his often belligerent and sarcastic approach caused his associates many problems. Liddell Hart perceived this at the time, and in 1929 he asked Swinton, then Chicele professor of military history at Oxford, whether there was any way out of this impasse. Swinton replied that sharp prodding by reformers was usually necessary to get action taken, but he also noted that such an approach was essentially negative and that one could "express strong convictions without going out of the way to aggravate other people or put their backs up." Sharp attack such as Fuller's, he added, might be understandable, but they might also easily become self-defeating in the end because, after all, it was the results that counted.[18] These criticisms contain considerable validity. While there is little doubt that the very content of the theory of armored warfare would have aroused a great deal of opposition within the army, it is important to note that some of its advocates were sufficiently alert to realize that there was no wisdom in deliberately antagonizing everyone. Fuller, on the other hand, never displayed any concern whatsoever about avoiding unnecessary antagonisms. It is this, far more than anything else, which made him so unsettling a cohort.

Seventeen years younger than Fuller, B. H. Liddell Hart was born on October 1, 1895, in Paris, where his father was serving as a clergyman. In 1901 his family returned to England, where he began his schooling and developed a strong interest in military history. According to his memoirs, it was about the age of thirteen that he first became aware of what he later perceived to be a national characteristic of the English people. A leading scorer of his school's cricket team, he was dropped because his manner of play was highly unorthodox, his effectiveness notwithstanding. It was, he later reminisced, "my first awakening to the national tendency in recent time to cherish Balaclavas and Dunkirks far more than victories," if the latter had to be won by "unorthodox means."[19]

Liddell Hart entered the army as a Cambridge candidate when the war broke out in 1914 and served as an infantry officer on the Western Front. Wounded on the Somme in 1916, he was placed on light duty and occupied himself by writing articles on infantry tactics. His ideas crystalized shortly after the war in a two-part concept of small-unit operations that he termed "The 'Man in the Dark' Theory of Infantry Tactics" and "The 'Expanding Torrent' System of Attack." In them, he compared combat to a struggle between two boxers fighting in the dark. With one arm outstetched to protect himself and locate and grasp his opponent, each man keeps his other arm held back, cocked for the decisive blow. In the same way, he reasoned, infantry tactics should be seen as a series of maneuvers between opposing forces, each one with a guarding element and a striking element. Although he did not see the potential of armored forces at this time, the emphasis on flexibility and maneuver that would become the hallmark of his later writings was obviously present.[20] These ideas were reflected in the postwar British training manuals he helped to write.

Liddell Hart joined the short-lived Army Education Corps in 1919 because of the continued disability resulting from his wounds. In 1922, he tried to join the Royal Tank Corps, enlisting the support of Fuller, but was unable to pass the Medical Board. In 1924, he was placed on half-pay and the following year joined the staff of the *Daily Telegraph,* replacing the recently deceased Charles A'Court Repington as the only full-time military correspondent in England. According to his agreement with the newspaper, Liddell Hart's duties were primarily analyzing rather than simply reporting military affairs. Consequently, he was able to use his post to comment on a wide variety of military policies and especially to urge the army to mechanize and expand its tank forces. His trenchant style and perceptive analysis quickly made him one of the most influential military writers of his day and often persona non grata at the War Office. In 1935, he moved to a similar position at the *Times* and two years later, he became the unofficial advisor to the new secretary of state for war, Leslie Hore Belisha, over whom he exercised a powerful influence for just over a year.

Like Fuller, Liddell Hart's writings cover an impressively broad scope. His first interest was the defense problems of his day, but he turned to history for many of his ideas and made important contributions in this field as well. His biographies of Scipio Africanus, William T. Sherman, Lawrence of Arabia, and Marshal Foch are still regarded as among the best, if not the best, on their subjects. Also like Fuller, he was a prolific writer. In addition to hundreds of news articles, he was military editor of the *Encyclopedia Brittanica,* contributed numerous articles to various military journals and periodicals, and wrote a total of nineteen books, some of which were collections of previously published articles.[21]

Personally, Liddell Hart was far more tolerant and broad-minded, and, one suspects, likeable than Fuller. He seems to have seen himself as both a scholar and a teacher, for he was not merely content with finding and stating the truth but with winning others to his cause. He expressed himself clearly and powerfully but without the savage sarcasm that so often characterized Fuller's writings. Consequently, while Liddell Hart was just as outspoken as Fuller, he never alienated people to the extent that Fuller did. His correspondence with Field Marshals Montgomery-Massingberd and Deverell, both of who occupied the post of chief of the imperial general staff in the 1930s, reflects a degree of respect even though there was often strong disagreement between the men. Liddell Hart also had a fairly close relationship with Field Marshal Ironside, CIGS in 1939 and 1940, who could describe him as "so hideously conceited . . . that he isn't nice to meet. A cad too," and then a few weeks later write, "I must get hold of Liddell Hart and talk to him. He is always interesting."[22]

Comparisons between Fuller and Liddell Hart are almost inevitable. The two first met in 1920, and, although they remained close for almost half a century, their relationship was stormy on more than one occasion because of differences of opinion and their strong personalities. On the broad themes they were in almost complete agreement, but they often differed in the manner of their approach and their points of emphasis, the details of which will be discussed shortly. In general, the most important distinction between their approach was that Fuller had a penchant for couching his arguments in philosophical terms while Liddell Hart employed a simpler, more direct approach. Whether or not this difference reflected a greater intellectual power on Fuller's part is questionable, however; indeed, in many respects the only difference between the two is that Liddell Hart is far easier to read. As to their respective contributions to the theory of armored war, it is clear that Fuller was its originator and first advocate, Liddell Hart not joining the ranks of the committed until the early 1920s. As late as 1919, in fact, Liddell Hart wrote that, while there was always constant need for change,

> we need not, however, let our imagination run riot in the manner of some popular novelists who pose as military critics, forecasting fleets of giant ironclad land ships and swarms of armoured aeroplanes [sic], whilst they consign the ordinary infantry man to the limbo of the past.[23]

Even as late as 1922, he wrote to Fuller, "I am an ardent believer in tanks and only want to be converted entirely. Full conviction is, however, necessary if one is to be a good advocate."[24] Yet to assume that Liddell Hart was merely Fuller's disciple is to misunderstand their relationship and to overlook the former's very real and highly important contribution to the theory of armored warfare. Their early correspondence, it is true, reflects a superior-subordinate relationship—Fuller, after all, was a full colonel and Liddell Hart only a captain—but by 1924, this had clearly changed. In the following year, Fuller wrote to Liddell Hart:

> Your opinion counts a great deal with me—remember it was your criticism which gave me eighteen months of hard work—twelve in rethinking and six in rewriting my Foundations, but don't make me write them again for a year or so.[25]

The correspondence between the two men is both voluminous and fascinating. It contains very little of a direct personal nature, being confined almost wholly to an exchange of ideas and views. At times, their disagreements were sharp, but never was there any loss of respect. In 1929, for example, both men were delving into the campaigns of the American Civil War, Fuller for his biography of Grant and Liddell Hart for his of Sherman. Disagreement had arisen over the origin of the successful Union strategy for 1864–65, each man arguing that his subject was the source. Finally, after summing up all the points in his favor, Liddell Hart told Fuller: "The trouble with you is that you have never been trained in any school of history to develop the 'historical point of view,' " The problem, he continued, was that "one so often has the feeling that you don't sift the evidence but simply search it for something to support an opinion already formed."[26] Such a challenge could not be ignored. Three days later, Fuller answered in a letter addressed to "Dear Lord God Almighty!" in which he declared, "I strike my head two and seventy times on the ground before your superior historical training"—which, as both men knew, consisted of a single year at Cambridge. Characteristically, he then proceeded to launch a spirited attack on his friend's arguments and conclusions.[27] Such was the relationship between the two men that it could withstand many such exchanges and yet flourish. In the end, their partnership—and insofar as the development of the strategy of armored war was concerned it was just that—nurtured both without costing either his individuality.

With Fuller and Liddell Hart were a small band of middle-grade officers within the army who faced the unenviable task of urging the acceptance of the new theories while, at the same time, working out the mechanics of their application. To many outsiders, there seemed something very sinister about this whole group, and many years later Field Marshal Ironside recalled:

> What irresponsible harm he [Liddell Hart] did by forming a sort of clique at his house with Pile, Hobart, Martel, Broad and company sitting, hoping for high commands. It makes me quite sick to think of it.[28]

Yet nothing could be further from the truth. Although none of these men achieved the stature of Fuller or Liddell Hart as military theorists, all were extremely capable and innovative soldiers of strong conviction, as, indeed, they would have had to be to promote a set of ideas so unpopular in certain influential quarters. They had a common purpose, but disagreements among them were not unknown, and, in the case of Martel and Hobart, these difficulties eventually developed into an intense dislike for each other.[29] Nor can self-seeking motives be ascribed to them. Given the army's attitude to Fuller and Liddell Hart, only a fool could have believed any association with them would advance his career. All would eventually achieve general-officer rank, but only one would be employed as a commander of tank forces in the field. They had the dispiriting experience of advocating the expansion of the army's tank forces during the interwar years, only to see others take command of those forces when they came into existence during World War II.

Of the five men in this small group, George M. Lindsay was the oldest. Born in 1880, he received a militia commission in 1898 and a regular commission in the Rifle Brigade two years later. He was an early advocate of the machine gun and a leading figure in the formation of the Machine Gun Corps in 1915. Later he helped develop the tactical principles governing the employment of that weapon. His first contact with armored forces occurred in 1921, when he was appointed to command an armored-car group in Iraq. Returning to England two years later, he transferred to the newly formed Royal Tank Corps and became chief instructor at the corps school. It was in this capacity and, later from 1925 to 1929, as inspector of the Royal Tanks Corps that Lindsay did his most valuable work for the tank forces. He was instrumental in the evolution of tank tactics and left his mark on virtually every aspect of the employment and support of the new weapon. After 1929, Lindsay's work with tanks was spasmodic—there were no tank commands higher than that of battalion—but he continued in the vanguard of army modernization until his retirement in 1939. He was often considered too kindly and sympathetic to be a truly effective commander, and, indeed, his ability in this regard was questioned even by his close friends. His poor showing in the 1934 maneuvers apparently confirmed this evaluation and was a major reason for the indifferent assignments he received afterwards.[30]

The next two men, Charles Broad and Frederick A. Pile, were both artillerymen. Broad was born in 1882 and entered the Royal Artillery from Pembroke College, Cambridge, in 1905. Like Lindsay, he did not come into direct contact with tanks until after the war. He performed his most effective work with the Tank Corps in the late 1920s, first as commander of the Tank Gunnery School and then, from 1927 to 1931, as the deputy director of staff duties at the War Office. During this latter period, he worked closely with Lindsay in writing the first manual on tank tactics and, in 1931, was given command of the temporarily formed First Brigade, Royal Tank

Corps. Not a dynamic personality, Broad was known as the "brain" among the tank advocates. He was a serious and thoughtful person who seldom impressed people at first—a decided handicap for any commander—but who compensated for this through his considerable abilities and knowledge as a trainer and organizer.[31] Frederick Pile, known to friends as "Tim" because of his small stature, was born two years later than Broad and, again, had no direct contact with tanks until well after the war. In 1927, he was given command of the light-tank battalion of the newly formed Experimental Mechanized Force and, by his own admission, pushed his tanks as far and as fast as he could simply because he did not know what else to do with them. When it quickly became evident that the opposing forces could not cope with these radical tactics. Pile became a convert to mobile armored warfare.[32] As assistant director of mechanization in the War Office from 1928 to 1932, he worked closely with Broad and Lindsay on tank problems, but, like them, he had no further direct contact with the Tank Corps afterwards. In 1932, he was assigned to command the Canal Brigade and, when he returned to England in 1936, was given the task of building the nation's antiaircraft defenses.

The last two tank advocates, Giffard Le Q. Martel and Percy Cleghorn Hobart, were both engineers who entered the army from Woolwich. Martel served with the Tank Corps in World War I, but he refused a transfer to the corps when it became a permanent formation in 1923. His interest in tanks was primarily technical, and, although he developed the one-man tank in 1926 and served as assistant director of mechanization on the eve of World War II, his influence on the development of tank doctrine was minimal. Hobart—he preferred simply to be called "Hobo" for rather obvious reasons—was at his best as a commander in the field. Though often difficult to get along with, he was resourceful and dynamic and regarded by many as the best natural leader of the tank advocates. His work with the First Tank Brigade from 1934 to 1938 was undoubtedly the high point in the development of British armored forces between the wars. His difficult personality, however, caused him many problems both before and during the war.[33]

An obvious question at this point is what made these men distinct from so many of their comrades who failed to perceive the full possibilities of the tank. It is a question far easier to ask than it is to answer. Neither their backgrounds nor their careers reveal any radical differences prior to their first contact with tanks. None was a cavalryman, but the cavalry's opposition to mechanization has been greatly overstated, and, as has been noted, Pile has recalled working with a number of extremely progressive cavalrymen at the War Office in the late 1920s.[34] The singular difference between these men and their more conservative fellow officers is that (with the exception of Liddell Hart) all served with tank forces at some point in their careers. It was this direct contact, coupled with their obvious ability and receptive minds, which marked their point of departure from the mainstream of army thought.

The Theory of Armored Warfare

The first and most crucial fact to remember about Fuller and Liddell Hart's theory of armored warfare is that it was far more than a new set of ideas formulated simply to maximize the capabilities of tanks. Rather, it was nothing less than a revolution in military thought, a comprehensive philosophy of war explicitly designed to destroy the theoretical and practical foundations of the strategy of attrition that grew out of the experience of the Napoleonic Wars and dominated the strategic doctrines of the postwar period. The main thrust of Fuller and Liddell Hart's criticism was that this model oversimplified and distorted the true principles of war; that, as a result of these distortions, Europe was plunged into an era of total war; and that the subsequent devastation of World War I was senseless and unnecessary. They believed the need for radical change was obvious and urgent, for, in Liddell Hart's words, "it is surely clear that any further war conducted on similar methods must mean the breakdown of Western civilization."[35]

To prevent such a tragedy, both men attacked not only the concept of total war, but also virtually every aspect of pre-1914 military doctrine. Liddell Hart, in particular, recognized that Clausewitz was not fully responsible for all the ideas ascribed to him, but he was severely critical of the Prussian's failure to guard against misinterpretation.[36] They both attacked Clausewitz's claim that the aim of military operations should be the total destruction of the enemy forces on the battlefield, for, as Liddell Hart pointed out:

> there are . . . plenty of negative examples to prove that the conquest of the main armed forces of the enemy is not synonymous with victory. History has no more complete victories than Cannae and Sedan, yet the one failed to bring Hannibal to his goal, and the other was only consummated when Paris fell several months after.[37]

Furthermore, they maintained that such emphasis could only lead to total war, which, in Liddell Hart's words, was "a war in which the conductor does not know when to stop. It implies that the end is pursued regardless of what lies beyond."[38] Clausewitz was also castigated for his emphasis on the dominance of numbers in battle. In one of his most purple passages, Liddell Hart dubbed him the "Mahdi [prophet] of mass and mutual massacre" and claimed "not merely stalemate, but massed suicide—more truly homicide—was the penalty of Clausewitz's theory of mass."[39] They also decried what they considered to be the simplistic view of "will" found in many prewar writings. The war had demonstrated that any attempt to destroy an enemy's morale by attacking his physical being was both costly and of dubious value. Assessing Foch's analysis of "will," Liddell Hart caustically remarked that the only thing the marshal forgot was that "a will can only be effective inside a living body."[40] Finally, they attacked the dogmatism of prewar thinking. Fuller held this to be the primary reason for the failure of generalship during the war and traced its roots to inadequate prewar thinking. He described his instructors at the Staff College in 1913–14 as merely "parts of a machine created to produce standardized thinking."[41] Little wonder that, when the war broke out, Great Britain's military leaders were unable to adapt to new conditions. In one of his bitterest attacks on World War I leadership, Fuller recalled that during the 1917 Flanders offensive he had believed that Haig and his advisors were little more than criminals seeking only to gain glory and preserve their reputations. Twenty years later, the best he could say of them was to admit that such charges were unfair—but only because he had sadly underestimated their stupidity and lack of imagination![42]

The task Fuller and Liddell Hart set for themselves was to correct these errors in military theory and practice. To do this, they first had to discover the true principles of war, "to produce true grand strategists to replace the colour-blind exponents of mass destruction who can only see 'red.'"[43] But both men believed that this was only the first step toward correcting the problem because the course of World War I was not just the product of false doctrines; it was also the product of the belligerents' military systems, specifically the nation in arms and its most obvious product, the mass-conscript army. Mass mobilization engendered propaganda and aroused passions that made the task of maintaining rational wartime objectives extremely difficult, if not impossible. It also created armies of such size that commanders became paralyzed by what Liddell Hart called an "indigestable mass of infantry," and found it impossible to maneuver their forces. Consumed by the logistical problems of these forces, they were reduced to the status of "quartermaster officers."[44] The acceptance of the strategy of attrition and bloody slaughter, then, was not just the result of doctrine: commanders were impelled to this course of action because the instrument at their disposal hardly admitted any alternative. As a result, to formulate new principles of war would be of little value if military forces continued to possess their own momentum capable of propelling belligerents into an uncontrollable pattern of escalation. But in order to eliminate mass conscript armies, it was necessary to do more than simply decry their wasteful and hazardous nature; it was also necessary to devise a more powerful military instrument that would be capable of defeating them without creating its own impetus toward total war. In the final analysis, then, Fuller and Liddell Hart's theory of armored warfare was designed as an alternative to total war and the strategy of attrition, and the mechanized army organized around the capabilities of tanks was the instrument for putting the new theory into practice. To appreciate how they arrived at this conclusion and to understand better the nature of the opposition they faced from the British Army, it is helpful to examine their ideas under four basic headings: the nature and object of war, the goal and principles of strategy, the place of tanks in future war, and the nature of generalship.

I

Although Fuller and Liddell Hart's view of the nature and purpose of war was not, for the most part, original, it did run counter to prevailing military doctrines. They believed that war was an integral part of human activity, and that it was, in essence, a struggle of wills; a struggle, they stressed, that could be fought on nonmilitary as well as military grounds. As Fuller explained in 1928, war was a struggle between two wills, each possessing economic resources, ethical codes, social rights, and military forces. Each of these elements could be attacked in its own way, singly or combined: economically by attacking commerce and industry; "morally" by attacking the will and nerves of the civil population; socially by attacking domestic institutions and tranquillity; and physically by attacking the military forces. Of the four, he concluded, the military approach is the least direct and the "moral," the most direct.[45] This assumption, prominent in the writings of both men, led them to advocate the strategic bombing of civilian population centers as a means of breaking the enemy's will. They were not, however, calling for the type of area bombing that occurred in World War II; they believed it would only require a minimal amount of physical destruction to produce such a disruption of normal life that the enemy would prefer peace to a continuation of the struggle, and both urged the use of gas as an even better alternative to high explosives.[46]

This view of the nature of war led both men to the conclusion that military victory was not an end in itself, but only a means to an end. In Fuller's words, the purpose of war "is not to kill soldiers or sink ships, but to change a policy which these ships and soldiers are protecting."[47] Liddell Hart expanded on the same theme in his first book on military theory, *Paris; or the Future of War.* Defining the purpose of government as insuring its citizens "an honorable, prosperous, and secure existence," he wrote that there can only be one legitimate purpose in war: "To ensure a resumption and progressive continuance of what may be termed the peace-time policy with the shortest and least costly interruption of the normal life of the country." He continued:

> What stands in the way of this? The determination of the hostile nation to enforce its contrary policy in defiance of our own aims and desires. To

> gain our aim or objective, we have to change this adverse will into a compliance with our own policy, and the sooner and more cheaply in lives and money we can do this, the better chance is there of a continuance of national prosperity in the widest sense.
>
> The aim of a nation in war is, therefore to subdue the enemy's will to resist with the least possible human and economic loss to oneself.[48]

There was little doubt in either man's mind that the political and military leaders of all the belligerents in World War I had failed to see this basic truth. In his studies of military history, however, Liddell Hart believed he had found an almost perfect example of the kind of leadership required in the person of Scipio Africanus, the Roman general who finally defeated Hannibal in the Second Punic War. In addition to praising the Roman's martial ability, Liddell Hart emphasized that Scipio understood both the interplay among military, political, and economic factors in war and the fact that military victory is only a means to an end. Furthermore, he held that Scipio applied this understanding in the peace that followed his victory over Hannibal at the battle of Zama. By the standards of the day, Carthage would have been totally destroyed, but Scipio showed true moderation. The final peace terms assured Rome's security, gave the Carthaginians no reason to seek revenge, and resulted in fifty years of peace, unspotted by Carthage and even producing an offer to aid Rome during a conflict in the eastern Mediterranian. No wonder Liddell Hart called Scipio "a greater than Napoleon": he not only knew how to win wars, but he also knew how to win the peace.[49]

Although Fuller's analysis of the nature and purpose of war reached essentially similar conclusions, his perspective was given a somewhat different orientation by his belief in economic determinism. Historically, he wrote, considerations such as security, religious freedom or domination, vengeance, and economic prosperity all had played a role at one time or another in causing war. He maintained, however, that since the rise of industrialism, the world order rested on economic foundations, and, since war was rooted in the societies involved, all modern wars, therefore, had to be economic in their origins and nature. Thus, World War I had not actually begun in 1914 and ended in 1918, but only had changed its form from economic to military and back again in those years. Unfortunately, national leaders had not recognized this fact and made no attempt to adjust their goals accordingly. As a result, what was essentially an economic struggle became an ideological struggle, in which all the participants lost far more than they could hope to gain.[50]

Fuller was equally pessimistic about the future, because he believed too few people understood these facts despite the experience of the war. He believed Europe's postwar difficulties were not a result of the war itself but of the peace treaties that came afterwards. Because they were supposedly based on moral rather than rational principles, these documents were dominated by a spirit of vengeance. Thus, while the war had caused tremendous physical destruction, Fuller was far more disturbed by the fact it had not destroyed enough as the "selfishness, greeds, fears and doubts" of the prewar era had been carried over to sow the seeds for future conflict.[51] At one point, he compared the post-Versailles era with the Reconstruction era in American history since, in both instances, the "bullying spirit" and "greed" consumed the victors. Rather than establishing a better peace for all, the victors, by setting out to punish the vanquished, had only produced results detrimental to everyone's interests. Unfortunately, where the American Civil War had at least resulted in union and economic prosperity, World War I only produced an even greater fragmentation of Europe thereby reducing even further the chances for an enduring peace.[52]

Just as both men found total war unacceptable, so too they found total peace—pacifism—unrealistic. While Liddell Hart's objections to pacifism rest on an essentially negative view of man, Fuller, consistent if nothing else, rested his opposition on economic grounds. War and weapons, he wrote, were merely symptoms of deeper stresses in the international system—essentially the competition for raw materials and markets—and could not be eliminated until statesmen attacked the root of the problem. Although no idealist, Fuller believed that war could be eliminated if nations realized they were no longer economically independent and that the prosperity of each depended to a significant extent upon the prosperity of all. Only if all nations could participate on a mutually profitable system, either through trade or economic federation, could the major cause of war be eliminated. Unfortunately, he saw little possibility of such an occurrence in the foreseeable future.[53]

The sum of Fuller and Liddell Hart's study of the nature and object of war, then, led both to the conclusion that total war constituted a major threat to Western civilization and that future wars must be limited if they were to serve the interests of the nations involved. The first step toward such a goal was creation of a new concept of strategy.

II

Although both men wrote prolifically on the subject of strategy, it was Liddell Hart who delved most deeply into the subject and developed the most comprehensive analysis. He began his study by describing the problems involved in definding the term.[54] The most important, he wrote, was that strategy existed in that delicate sphere of policymaking between political and military decisions, and that its ultimate control therefore depended on its definition. In a monarchical or authoritarian form of government, in which political and military leadership was vested in a single person, this problem was largely theoretical; but under a constitutional form of government, it was very real. If civilians secured too strong a voice, they might find themselves planning military operations for which they had no background. Worse, if military men assumed too strong a position, they might dominate purely political matters and the objective of the war would become lost in the efforts to achieve victory. A balance, therefore, had to be struck that allowed both civilian and military leaders to fulfill their responsibilities without interfering with each other.

Liddell Hart dealt with this problem by first distinguishing between two terms closely related to strategy: grand strategy and tactics. The first was almost synonymous with policy, its purpose being "to coordinate and direct all the resources of a nation toward the attainment of the political object of war—the goal defined by national policy." Located at the other end of the spectrum, tactics was defined as the disposition and control of troops once contact had been made with the enemy.[55] Since the first of these terms was clearly political and therefore the responsibility of civil authorities, and the second was clearly military and therefore the responsibility of the armed forces, neither presented a problem. The gap between the two had to be bridged, however, by coordinating them through a process of logical planning. This was the purpose of strategy, and Liddell Hart's definition of the term bridged this gap. It was, he explained, "the distribution and transmission of military means to fulfill the needs of policy."[56] This definition clearly subordinated strategy to policy and, by implication (which he stated later), fixed the responsibility of both the military and the civilian elements in government. It gave the military control over the execution of strategy, but withheld from them control of its actual formulation. The latter, he felt, should be the ultimate responsibility of civilians as too many nonmilitary considerations were involved. A decision to hold one area and allow the enemy to seize another, for example, had to take into account political, economic, and even psychological as well as military considerations. Commanders who opposed a strategic decision were given no alternative other than resignation; they must not determine strategy or change it once a decision was made. This was the prerogative of the government alone.

Liddell Hart next considered the principles that should govern the execution of strategy.[57] He began by postulating the thesis that the objective of strategy, in accordance with his description of the purpose of war, was to "diminish the possibility of resistance," and since resistance occurred mainly in battle, strategy should aim to reduce the possibility of fighting to the barest minimum consistent with the objectives and limits established by the government. The ideal solution was the destruction of the enemy forces without battle, as Turenne had done in his last campaign in Alsace or Napoleon did at Ulm, but Liddell Hart realized that such opportunities would rarely arise. If a battle has to be fought, strategy should aim to create a situation in which the enemy could be defeated at the least cost to oneself.

How could this be done? Liddell Hart's answer lay in what he termed the principle of the indirect approach. In essence, the indirect approach was a strategy of dislocation through paralysis rather than annihilation through attrition; its aim was not to overwhelm the enemy by employing strengh against strength, as had been attempted in World War I, but to immobilize and confuse him by striking so as to upset his mental and physical equilibrium. It avoided a direct confrontation on the grounds that it was excessively costly and reduced rather than enhanced the chances of success because the enemy would be ready to meet the blow. In practice, he further explained, the indirect approach involved operations such as turning maneuvers, which forced the enemy to make a sudden change in front, movements that threatened the enemy's supply lines, or advances that threatened two or more objectives simultaneously. In any event, movements should be along the lines of least resistance and expectation so that the enemy would be unable to concentrate his strength at a single point. In Liddell Hart's view, this concept was not only superior to the direct approach, but also fitted in perfectly with the need to reduce the destructiveness of war. No military operation could be totally devoid of casualties, as he well recognized; but by emphasizing mobility and maneuver over mass and attrition, it could reasonably be expected that a significant reduction in losses could be achieved.[58]

Liddell Hart's greatest work of military theory, *The Decisive Wars of History*,[59] is a study of this principle based on historical analysis. He began by comparing the indirect approach with the direct and then surveyed military history from the ancient Greek wars through World War I to determine how each had fared. His conclusion, based on a study of nine wars of ancient history and eighteen major wars of modern history, a total of 240 campaigns, was that the indirect approach had the overwhelming advantage. He noted that, of all the campaigns covered, only six were won by the direct approach: Alexander the Great at Issus and Gaugemela, Napoleon at Friedland and Wagram, and the elder Moltke at Sadowa and Sedan. Furthermore, he claimed, there were additional considerations that made it unlikely that these campaigns owed their success in any way to the direct approach: in the first two, the grand strategy employed the indirect approach and Alexander had a far better army than did the Persians; in the second two, Napoleon began the campaigns with the indirect approach and even though he was victorious, his casualties were heavy; and, in the final two, the Germans had superior weapons in the first and superior numbers in the second (not to mention an unbelievably incompetent enemy), both of which were vital elements in the final victory.[60]

Despite this, however, the direct approach had been and was in his own day the accepted form of strategy. The indirect approach was used only when an army was in a greatly weakened position, thus underscoring the validity of this argument. Liddell Hart summarized his position with two maxims drawn from this study of military history:

> [First] in the face of the overwhelming evidence of history, no general is justified in launching his troops to a direct attack upon an enemy firmly in position. The second, that instead of seeking to upset the enemy's equilibrium by one's attack, it must be upset before a real attack is, or can be successfully, launched.[61]

III

Given these definitions and governing principles of the nature and objective of war and strategy, it is now possible to describe the instrument through which Fuller and Liddell hoped to implement their ideas: the mechanized army organized around the tank. In their view, such an army would possess two overwhelming advantages. The first was that it would be far superior militarily to the armies of World War I in every way. It could cover greater distances in far less time, producing what was, in essence, a concentration of space and time. It could combine firepower, shock action, and protection to a degree hitherto impossible and achieve this with greater mobility and fewer troops. The armies of World War I admittedly had tremendous firepower and could delivery powerful blows against an enemy, provided he was pinned down. But, Fuller and Liddell Hart asserted, they could never accomplish this against a mechanized force, which could dodge their blows while striking at vulnerable points in their defenses. By the time the enemy reacted, the mechanized force would be on its way to another area. Finally, a mechanized army could deal more effectively with airplanes and gas, two weapons that had figured prominently in World War I. The old-style armies were unable to neutralize either of these weapons without further impairing their effectiveness: against planes, they could do little except run for protection, and against gas, they could only wear masks, which further restricted their movement. They also required an enormous administrative and logistical echelon, which only added to their vulnerability to gas and air attack. A mechanized army, on the other hand, would go a long way toward solving these problems. Its tanks could be sealed and, with their armored shell, be relatively safe from both gas and planes. Because it would be smaller and therefore require a smaller support echelon, it would offer a smaller target. Finally, its cross-country mobility would enable it to disperse more quickly and effectively. In short, Fuller and Liddell Hart believed mass-conscript armies were being challenged by a number of new weapons with which they could not cope. It was obvious to them that the only way nations could hope to survive in modern war was to scrap these armies in favor of completely mechanized forces built around the tank.[62]

The second advantage of the mechanized army was that it would assist in the necessary effort to limit the modern tendency toward total war. The problem with mass-conscript armies was that they required mobilization of the entire nation, which led to total war. A mechanized army, on the other hand, would not require mass mobilization because it could not exist on a mass-conscript base. The cost of mechanization would compel a reduction in the size of armies, and the training requirements of the new forces would be so much greater that the troops' minimum length of service would have to be increased. Large numbers of short-term, conscripted troops would have to give way, therefore, to smaller numbers of more highly trained men. The natural extension of this development was a return to the relatively small, professional armies characteristic of eighteenth-century Europe, and both men, in fact, reached this conclusion. By returning to the military institutions of this earlier period, they believed a major step would be taken toward the goal of restoring limits and a degree of rationality to war.[63]

Neither Fuller nor Liddell Hart, however, relied solely on military arguments to support their position. Fuller, in particular, argued that, since a nation's military forces were a reflection of its society, it was becoming increasingly anachronistic for the British Army to continue to rely on muscle power. The old-style infantry and cavalry were products of preindustrial societies; with the rise of modern industry, it was only logical that the army make a similar transformation. Railroads and motor cars had made a significant impression on military operations, but they were "administrative" means of movement—they could move men and supplies to the battle area, but during actual combat they were of little tactical value. In Fuller's words, the ultimate effect of mechanization was to endow "the army with a higher energy than it now possesses today."[64] The process of mechanization had begun; it was now time to finish it.

Having established the value of mechanized forces, Fuller and Liddell Hart then turned to formulating a set of principles to direct them. The source of their ideas is found in the plan drawn up by Fuller in May 1918 aimed at ending the war against Germany in the following spring. Entitled "The Tactics of the Attack as Affected by the Speed and Circuit of the Medium D Tank," or, more commonly, "Plan 1919," it was essentially an adaptation of the tactical methods employed by the Germans during their offensives of that spring.[65] In those assaults, the Germans had relied primarily on select combat groups to filter through the weak points in the Allied lines to isolate forward units from their headquarters. Destruction of the British command system rather than destruction of troops had been their objective. During visits to the front at the time of those attacks, Fuller had been struck by the tactical successes of this system. Isolated units quickly became disorganized even though they had been little affected by direct contact with the enemy. Many had lost all sense of direction and had become totally ineffective as fighting units. Fuller's plan was identical in purpose to the German system, with the exception that he gave the mission of penetration and disruption to tanks instead of infantry.

Plan 1919, which is as much a tactical study as it is a plan of operations, opened with a brief description of the characteristics and advantages of tanks. Among them, Fuller cited increased mobility, protection, and offensive power, and he asserted that tanks could be put to best advantage by allowing them to operate independently, that is, not limited by the pace or tactics of infantry. This change, Fuller admitted, was a radical departure from previous tactical methods, including those he had devised himself for the battle of Cambrai when he had employed tanks as little more than a battering ram to pave the way for advancing infantry. But, he explained, these tactics had been the combined product of inexperience with tanks and the extremely primitive state of the weapon itself. As the Mark IV's had a maximum speed of only 3.7 miles an hour under the best conditions, the question of releasing them from cooperation with the infantry was only theoretical. Now, he reported, a new tank was being designed capable of radically changing this situation. Designated the Medium D tank, it would be able to travel at least twenty miles an hour for 150 to 200 miles, cross a thirteen-foot gap, and be light enough to cross ordinary road, river, and canal bridges. Although the tank was only in the early stages of development, Fuller expressed confidence that a sufficient number could be produced by the following spring to carry out a major operation, and he proposed a new tactical system to maximize its capabilities.

The first step in Fuller's plan was for the Allies to force or bait the Germans into concentrating their reserves in certain areas where they would not be in a position to counter the main Allied thrusts. When this was accomplished, the Allies would strike at four narrow sectors along a ninety-mile front—an area Fuller termed the "morcellated front of attack." Each attack would be launched in three stages, conducted by separate waves: first, a breaking force composed of heavy tanks would clear a path through the enemy's defensive zone; second, a "disorganizing" force composed of the Medium D tanks would race to knock out each German division, corps, and army headquarters in its sector; and, third, a pursuit force composed of all types of medium tanks would complete the disorganization and destruction of the German forces. The key to the success of this plan lay, obviously, in the second stage. Fuller called for the location of as many German headquarters units as possible and the assignment of a specific force of tanks to each one. Once these objectives had been obtained, the success of the plan was guaranteed: the German Army would be paralyzed, the cohesiveness and fighting effectiveness of its units would be destroyed, and the mass of soldiers would become as useless as the limbs of an individual who had been "shot through the brain." Fuller assigned a vital role to aircraft throughout the attack. They would be the advance guard for the tanks, acting as scouts and airborne artillery. Infantry, on the other hand, would play only a minor role. They were assigned only four tasks, none of which was of major importance: to assist in the initial tactical penetration, to operate in areas unsuited to tanks (and since the attack sector would be chosen on the basis of its suitability to tank operations, this function would be small, indeed), to occupy captured terrain, and to support the logistical and administrative services.

Unfortunately for the later development of British armor, Plan 1919 was stillborn. What would have happened had it been put into effect can only be speculated upon, but it appears likely that the plan had an excellent chance of being adopted, for the imperial general staff in London approved it on July 20, 1918 and forwarded it to Foch, who also agreed to its feasibility.[66] Certainly the development of the tank would have taken a far different course had the plan been put into operation.

Plan 1919 contains the crux of all of Fuller and Liddell Hart's early writings on the employment of armor. The plan itself was expanded into a general theory of war,[67] and its principles formed the basis for much of their later writings. Fuller's phrase, "shot through the brain," describes its essence perfectly. The enemy's command system and not the masses of enemy troops was its target. Either by a physical assault by tanks and aircraft or by an indirect psychological assault, its purpose was to disrupt coordination between commanders and their men. In either case, both Fuller and Liddell Hart believed that the use of tanks in this manner was the only way to avoid a repetition of the mass slaughter of World War I. In essence, they reduced the target of military operations from millions of troops to a relatively few key personnel; and while they obviously were too realistic to think that casualties could be restricted solely to this group, they believed that the scale of destruction could be measurably reduced by so shifting the target and that more reasonable and secure peace settlements could be concluded. Fuller summarized his system in one of his most descriptive passages:

> An army is an organism, comparable, like all other organisms, very closely with the human body. It possesses a body and a brain; its fighting troops are the former, its headquarters staffs the latter. In the past the usual process of tactics has been to wage body warfare: One body is moved up against the other body and like two boxers they pummel each other until one is knocked out. But suppose that boxer "A" could by some simple operation paralyze the brain of boxer "B," what use would all boxer "B's" muscular strength be to him, even if it rivaled that of Samson and Goliath combined? No use at all, as David proved!
>
> Now apply this process to the battle of 1923. The tank fleets, under the cover of dense clouds of smoke, or at nighttime, move forward, not against the body of the enemy's army but against his brains; their objectives are not the enemy's infantry or the enemy's guns, not positions or tactical localities, but the billets of the General Headquarters staffs—the Army, Corps, and Divisional headquarters. These they capture, destroy or disperse; what then is the body going to do, for its brain is paralyzed.[68]

Logically, this concept could ultimately lead to the total abolition of infantry, and some tank advocates did, in fact, reach this conclusion, much to the detriment of their cause. This so-called all-tank school argued that the foot soldier was simply a creature of the past who would be of no value whatsoever on future battlefields. In what was probably the first expression of this doctrine, "A Tank Army," written in November 1916, Gifford Martel had argued that future war would be fought between masses of tanks maneuvering over the landscape much as ships traverse the sea. Different types of tanks would take over the roles of the three traditional combat arms, which would all but disappear.[69] Fuller expressed this same conclusion in his Gold Medal essay. Urging the reader to bear in mind "that specialist corps, such as machine-gun, tank, or infantry, do not tend towards efficient cooperation," he argued tanks could conduct operations more efficiently and less dangerously than infantry and concluded, "Tanks, therefore, can replace infantry," and should largely replace or absorb all of the other branches of the army.[70] Even Liddell Hart, who in 1921 had decried those who questioned the importance of infantry, briefly accepted the total demise of this former branch of service at the time of Martel's development of a one-man tank in 1926. Writing of the new machine in the *Daily Telegraph* in March of that year, he claimed "It promises to reintroduce on the battlefields of the twentieth century the medieval knights in armour—modernized with caterpillar track legs and a petrol engine heart."[71] It was possible, he continued, to build an entire army around the mobile arm much as the Mongols had done in the thirteenth century, and, shortly after, enthusiastically pictured a battlefield where 60,000 Martel tanks utterly routed a far larger opposing force organized and equipped along traditional lines.[72]

For the most part, however, Liddell Hart and, to a lesser extent, Fuller displayed more perceptiveness than this regarding the future of infantry. Both, to be sure, claimed that infantry had ceased to be the decisive arm on the battlefield, but they conceded it still had an important if limited role to play. Liddell Hart even came remarkably close to the idea of the tank-infantry team that proved so effective in World War II. In *Paris; or the Future of War,* he suggested that infantry might be of considerable value if they rode into battle on tanks and dismounted to clear out enemy strong points, adding that such troops should be called "tank marines" to denote the distinction between themselves and traditional infantry.[73] Fuller, on the other hand, never departed far from the position he established in Plan 1919: on the offensive, infantry lacked the necessary firepower and protection to operate effectively on the modern battlefield and, therefore, were only suitable for clearing up rear areas and guarding vital points. He admitted they might be of value in areas such as heavy woods, swamps, or very mountainous country, calling such places "tank-proof areas," but he argued such conditions were seldom found in decisive areas of operations.[74]

On the defensive, however, both men eventually saw a far greater place for infantry. Every army operates from a defensive base whose role—to hold the enemy in place while the decisive attack is made—requires little mobility but great firepower. This role, although less glamorous than that of the forces making the decisive attack, is equally essential, as it prevents the enemy from shifting his position to avoid the attack or even from returning a blow at the maneuvering force while it is on the move and vulnerable. In history, this has been traditionally the role of infantry, and both Fuller and Liddell Hart believed this should not be changed. But they strongly opposed the postwar organization and equipment of the British infantry because they felt these forces were totally inadequate to fulfill this role on the modern battlefield. Instead, they believed that the ideal infantry of the future should be patterned on the principles established by General Sir John Moore for light infantry during the Napoleonic Wars.[75] They would be highly trained elite troops, especially skilled in marksmanship, infiltration, the use of cover and concealment, and individual initiative. Although their mission would be defensive, they would conduct a highly active, rather than a passive defense, constantly harassing the enemy to keep him occupied and assisting in the destruction of his equilibrium. Finally, because the training requirements of such troops would be so great, they, like the tank crews, would have to be long-term professionals rather than conscripts.[76]

Considering the criteria set forth for both tank and infantry operations, it is clear that Fuller and Liddell Hart's ideas were not so revolutionary as they were often perceived. This impression is reinforced by their numerous references to earlier military organizations and methods of operations. The weapons they advocated were, indeed, revolutionary, but, as they constantly maintained, the principles of war never change. What was wrong with the existing method of waging war was that these principles were no longer being followed. Blinded by the glare of Napoleon's victories and overwhelmed by masses of infantry, the art of war had degenerated into a senseless slugging match; Fuller and Liddell Hart were simply trying to restore sound principles that had been forgotten. In this historical perspective, it can be said that both men were, in fact, reactionaries. It was their instruments that were revolutionary, not their goals. Thus the tank, which was to replace the cavalry, was heralded by Liddell Hart as merely a new means of bringing about "the long-awaited rebirth of the cavalry charge."[77]

Nowhere is this historically reactionary element more apparent than in Fuller's magnum opus on armored warfare, *Lectures on F. S. R. III.*[78] In this work, he looked beyond battles between mechanized and mass armies, which he believed represented only the first stage of transition to the new form of war. Once the superiority of armored warfare had been demonstrated, all armies would be forced to change in order to survive and, ultimately all armies would be small, professional, and highly mobile mechanized forces. The purpose of this book was to establish the operational guidelines for this future period.

Fuller opened his analysis with a description of the principles on which these future armies would be organized, principles frankly patterned after the model established by Alexander the Great over 2,000 years before. The Macedonian conqueror had organized his army into two elements: a defensive base that was the phalanx and an offensive wing that was the heavy cavalry. Likewise, Fuller organized his model army into a defensive antitank base and an offensive tank force. The antitank force included, in addition to infantry and antitank weapons, supply and administrative personnel providing support for the entire force. The offensive wing included scout vehicles and mechanized artillery, but was organized around the medium tank. All elements were so highly mechanized that the defensive base could move with the offensive wing prior to contact with the enemy, determining their ultimate position only when battle was imminent.[79]

Fuller next described in great detail the operations of these forces both on the offense and the defense. His vision of the operations bears a striking resemblance to those of the eighteenth century, concentrating on maneuvers to isolate the enemy from his supplies and attacks from unexpected quarters to upset his equilibrium. His emphasis, in line with the military theories of that earlier period, was clearly on maneuver rather than battle.[80] These traditional aspects of Fuller's theories were matched, however, by the revolutionary means through which he hoped to achieve them. The infantry was relegated to a very minor role, the cavalry was totally displaced by tanks, and aircraft were given a major role. In fact, he considered tanks and aircraft to be mutually dependent, for without aircraft to provide long-range reconnaissance the tanks would be blind and without tanks the air bases would be unprotected. "For a long time to come," he commented, "the one is unlikely to be able to operate safely without the other."[81]

Armored Warfare reveals Fuller at his best. It also demonstrates most clearly what he and Liddell Hart were trying to achieve with the strategy of armored war. Tanks were not simply a way to win wars; rather, they were a way in which some sort of rationality could be restored to the conduct of war. They understood that, while tanks were an offensive weapon, victory depended on the judicious use of both the offense and defense and the cooperation of all combat arms. Mechanization was only a means of preventing the absurd waste that characterized the strategy of attrition. Skill and flexibility always had been the keys to victory in war; mechanization was merely the means whereby they would be restored to their proper place.[82]

These concepts were of limited value, however, unless accompanied, first, by an orderly plan for the transformation of the army from a mass-conscript force to a mechanized force and, second, by a detailed understanding of how the new army should be organized. These tasks were undertaken by Fuller. He envisioned the process of mechanization in three stages: first, the supply and administrative echelon; second, the artillery; and third, the tactical forces. This process was based on the logical assumption that, until the support elements of the army were mechanized, it would be useless to mechanize the fighting elements. Tanks that could travel fifty miles or more a day would be of little value if their support could only move ten. At the same time, increasing the mobility of the support elements would allow a decrease in their numbers, enabling them to do their job better and more efficiently and thereby leaving more money for mechanization.[83]

The detailed organization of a fully mechanized army was first drawn up by Fuller in 1928.[84] It rested on the principle that to operate effectively, an army must be capable of carrying out three functions: to find the enemy, to hold him, and to destroy him. To accomplish each of these tasks, Fuller called for the formation of three types of divisions. The task of locating the enemy was assigned to "Light Divisions," highly mobile forces in which nonmechanized and support elements were kept at a minimum. These divisions were composed of two brigades, each composed of one battalion of forty scout tanks and three battalions each of forty destroyer tanks. Each tank was assigned a spare crew to protect the administrative services of the division when not required as replacements for tank-crew casualties. The total strength of such a division was estimated at 5,600 officers and men. The holding force of the army, which he called its "trunk," was composed of "Holding Divisions." These divisions, although still far more mobile than anything in existence, required great staying power to accomplish their mission and would have to face the likelihood of frequently fighting in areas where tanks could not operate. Fuller concluded, therefore, that they should include an infantry force. The divisions were composed of two heavy brigades, each composed of one battalion of scout tanks, one battalion of heavy tanks, and two battalions of destroyer tanks. They would also have two brigades of self-propelled (or tracked) artillery for support and two aircraft squadrons for reconnaissance and protection from enemy aircraft. If cavalry could be so organized that it could travel fifty miles a day for two to four consecutive days, it should also be included for local reconnaissance, communications, traffic control, and police work. The divisions' infantry compliment was to be taken from the spare tank crews, each tank having a total of four crews. They would be trained as light infantry as well as tank crewmen, and during normal movement they would be carried in cross-country vehicles. Thus organized, Fuller believed such divisions could operate over all types terrain and under virtually any conceivable condition. Their strength, including supply and administrative services, would be about 14,000 officers and men. "Pursuit Divisions" were the last element of Fuller's new army. Because their mission required even greater mobility than the "Light Divisions," he organized their combat elements so they could operate without concern for their support echelon. The main combat force of these divisions was contained in two brigades each of three battalions. Each battalion was equipped with forty highly mobile pursuit tanks with two crews per machine, the latter being carried in cross-country vehicles so they could remain close to the tanks and replace exhausted or injured personnel with a minimum loss of time. Protection for the support echelon was provided by a battalion of destroyer tanks specifically designed for this task. The total strength of such a division would be about 4,200 officers and enlisted men.

In conclusion, Fuller estimated that a British army composed of two divisions of each type would be more than adequate to meet both imperial needs and any emergency on the Continent. Although the force totaled only 2,000 tanks and scout vehicles, and some 60,000 personnel, its combat power would be far higher than that of the much larger expeditionary force dispatched in 1914. Indeed, given the condition of the other armies of Europe at the moment, he asserted that the presence of such a force in a theater of war would probably be decisive.

IV

Having established the basis for a new and more rational form of war and proposed an instrument to implement it, Fuller and Liddell Hart then turned to the problem of reviving the art of generalship. Both men were convinced that the strategy of armored war would enable skill to dominate battlefields once again, but they also believed that there would have to be a new type of general capable of taking advantage of the proposed system. Thus, the new concept of war and the revival of the art of generalship were both essential for success; one could not succeed without the other.

As one might expect from his authoritarian nature, Fuller was a strong believer in the idea of the "hero in history." In his *Memoirs,* he quoted approvingly Thomas Carlyle's statement that "the history of what man has accomplished in this world, is at bottom the History of the Great Men who have worked here."[85] Accordingly, his ideal general was a dynamic personal leader who was not "a mere prompter in the wings of the stage of war, but a participant in the mighty drama."[86] Such men had dominated most of military history, but with the advent of conscription the situation was altered radically. He compared this change to that wrought by the Industrial Revolution; just as the latter had ended any close personal ties between employers and employees, the rise of the conscript army had eliminated personal relationships between commanders and troops and thereby destroyed one of the most essential elements of successful generalship.

According to Fuller, the absence of such ties between generals and troops was one of the greatest problems of World War I. Located far to the rear, commanders lived in a totally different world from their men and were often unaware of the true situation at the front. Their isolation was compounded by the fact that they remained to the rear and rarely even visited the front. This was not because they lacked courage, Fuller held, but because they believed they should stay to the rear so they could see the whole picture and effectively control the masses under their command. As a result, they had ceased to be true generals and became instead mere "office soldiers" and "telephone operators." Summarizing his feelings, Fuller wrote, "An army cannot be led by a phantom sitting in the earth ten miles to the rear. Leadership demands personal contact."[87]

Although Liddell Hart did not go so far as Fuller in stressing the importance of personal leadership, he closely agreed with him on most other aspects of the nature of generalship. This was most apparent in their comments on the need for flexibility of thought among senior officers. Both saw conservative thinking in the military as one of the most dangerous attributes of commanders and, unfortunately, as Fuller noted in his *Memoirs,* one of the most common:

> The soldier is the most conservative creature on earth. It is really dangerous to give him an idea, because he will not adopt it until it is obsolete, and then will not abandon it until it has nearly destroyed him.[88]

In agreeing with Fuller on this point, Liddell Hart asserted that he considered this to be the key to the art of generalship. In *The Ghost of Napoleon,* he wrote:

> Creative thought has often counted more than courage; far more even than gifted leadership. It is a romantic habit to ascribe to a flash of inspiration in battle, what more truly has been due to seeds long sown—to the previous developments of some new military practice by the victors, or to avoidable decay in the military practice of the losers.[89]

Fuller and Liddell Hart also agreed on the causes of military conservatism: the failure to study military problems scientifically and an emphasis on technique rather than art. These failures, they held, were apparent in many ways but nowhere more so than in the study of military history. Both believed strongly in the value of this subject, but they were equally adamant that the manner in which it was then studied in the British Army rendered it useless and even dangerous. In the first place, the army concentrated on exhaustive examinations of only a few recent campaigns, producing an accumulation of minutiae that not only overwhelmed students, but also made it impossible for them to see the broad sweep of history necessary for understanding the principles of war and the development of the military profession. Even within this limited scope, moreover, the study was very poor, for, as Liddell Hart noted, soldiers too often "treated their history as a sentimental treasure rather than a field of scientific research."[90] Regimental histories, in particular, were not designed to instruct readers in the conduct of war, but to perpetuate pride through tales of heroism and glory. They taught smugness rather than critical thinking and produced an attitude that made resistance to change inevitable.

In their criticism of the methods used to teach future military leaders, Fuller and Liddell Hart then committed the gross heresy of concluding that, for positions of high command, long years of preparation, training, and working one's way up through the ranks were not only unnecessary but might well be detrimental! In his study of generalship, Fuller listed what he called the "three pillars of generalship" and purposely omitted any reference to long years of professional experience and training. His elements—courage, creative intelligence, and good health—were certainly important, but none of them depended in any way upon time or experience in the army. Liddell Hart went even farther and explicitly declared that professional experience was not necessary. Writing in the *Army Quarterly* under the pseudonym of "Bardel," he stated that the intelligent and serious amateur had, in fact, a distinct advantage over the professional because his view of war remained wider and his senses were not dulled by routine.[91] He specifically questioned the validity of military academies as well when, in describing the narrow curriculum of West Point in his biography of William T. Sherman, he asked whether great generals "have become great because or in spite of the system."[92]

To substantiate their thesis that longevity of service was not directly proportionate to military performance, both men employed a series of historical studies to determine the age at which outstanding commanders had achieved their greatest victories. In a study of 100 generals, Fuller noted their average age at the peak of their careers was 40.36 years and that the great majority were between the ages of thirty and forty-nine.[93] In a more restricted study, Liddell Hart chose fifteen men and, noting that eleven had reached their peak performance before the age of forty, concluded:

> Thus, in sum, the verdict of history is decisively against the contention that either age or practical experience of regimented routine is an aid to a mastery of strategy and tactics. Youth and independent study come well out of the argument with seniority and "practical experience," which too often have been synonyms for senility and mental atrophy.[94]

Both men believed that mechanization would greatly assist the process of bringing younger men to the fore because the requirements of mobile war would place too much of a strain on commanders for their positions to be held by any other than young men. But they also felt additional measures were necessary. First, they argued that men past their midforties should be prohibited from holding active field commands.[95] Second, and much more important, they advocated new methods of officer education. Broader experience was the key to their program, Liddell Hart going so far as to advocate abolition of the military academies and education of officers at civilian universities where they would receive a broader education. He also urged the creation of a special staff corps to be composed of outstanding university graduates specially trained for the purpose of high command at a relatively young age.[96] By such means, Fuller and Liddell Hart hoped to produce a new breed of military commanders, who would be qualified to employ effectively the strategy of armored war as they envisaged it and in so doing to restore the art of generalship.

The Theory of Armored Warfare and the Strategy of Attrition

Any assessment of Fuller and Liddell Hart's writings cannot avoid being influenced by the German victories early in World War II and the Israeli victories of 1956 and 1967. In each of these instances, the victorious generals relied heavily on fast-moving armored columns operating on principles borrowed, as they have freely admitted, from the two Englishmen.[97] Their successes were striking proof that Fuller and Liddell Hart had been correct in their analysis of the potential of armor and correct in the principles they established for its employment. Indeed, the success of the Germans and Israelis in paralyzing their opponents was so great that on occasion they came close to achieving the ideal of destroying the enemy's army without having to fight a major battle in the traditional sense. This is not to say that Fuller and Liddell Hart were perfect; it is to say that in the very risky field of prophecy, they achieved an enviable record.

Undoubtedly their greatest miscalculation in the field of tank operations was their misperception of the future of infantry. World War I had provided a number of striking examples demonstrating the vulnerability of tanks attacking without infantry, for in the absence of someone on the outside directing them, the vision of the tank crews was so limited that they can be subjected to withering fire with little opportunity to retaliate.[98] Liddell Hart, despite his brief infatuation with Martel's one-man tank, realized this, but he was unable to turn his fellow tank advocates from the "all tank" concept and convince the army that there was an alternative between their own position and that of the extreme tank people. He was undoubtedly hampered in this effort by the fact that Fuller, always one of the most visible of the tank advocates, never believed that infantry was of more than marginal importance, useful on the defensive, but an encumbrance anywhere else. This is particularly striking in Fuller's plan for the organization of a mechanized army drawn up in 1928. In calling for the infantry to double as spare crews for the tanks, with priority going to the latter role, he ignored the very real possibility that these troops might be heavily engaged as infantry when the call for help came. To withdraw them at such a time could easily court disaster either by creating a greatly weakened front or by stripping the force of a needed reserve. Furthermore, his proposal that these troops be highly trained as both infantry and tank crews required they attain a double set of skills that few troops could be expected to achieve. Much of Fuller and Liddell Hart's attitude toward the infantry was undoubtedly a by-product of their frustration over the army's refusal to see tank-infantry cooperation on any terms other than those employed at the battle of Cambrai. Nonetheless, their attitude did much harm to their cause, and proposals such as those contained in Fuller's 1928 plan for the organization of a mechanized army can only be described as a remarkable lapse for a soldier of his long experience.

As prophetic as Fuller and Liddell Hart proved to be in perceiving the potential impact of the theory of armored warfare on operations between armies, however, the fact remains that their ultimate justification for the adoption of this doctrine lay in their belief that it would inaugurate a return to limited wars. This, quite obviously, has not occurred. Part of the reason for this is that their concept of war has not been fully accepted by major warring powers (with the possible exception of Israel in the wars just noted). A deeper reason, however, is that Fuller and Liddell Hart failed to understand fully the origins and nature of total war and consequently expected too much from their own efforts to eliminate it. Both, to be sure, understood that a close relationship existed between a nation's ability to conduct war and its civil institutions and values. Both also understood that war could be waged on nonmilitary as well as military planes. But their arguments that total war was the product of misunderstandings regarding the true nature and objective of war, or of democracy, or of the adoption of mass-conscript armies were serious oversimplifications. The fact is, they misunderstood the extent to which total war was the consequence of the ability of modern, sophisticated governments to harness the wealth and productive capacity of an industrialized nation to the conduct of war. As a result, they failed to foresee the possibility either of mass-conscript mechanized armies or of the resilience of modern nations to withstand massive efforts to disrupt their capacity to wage war. There was a contradiction between Liddell Hart's claim that history had demonstrated that decisive battlefield victories did not necessarily lead to peace and both men's implicit belief that the destruction of an army through armored warfare could. In truth, the ability of a nation to wage war in the face of the destruction of its army does not depend on how that army was defeated so much as it does on the ability of the nation to devise a new means of resistance. The French defeats in 1870 and 1940 are cases in point. In both instances, the French armies were virtually destroyed within a few weeks of the opening of the campaigns. In the first, however, French resistance continued for some months, while in the latter, in which blitzkrieg strategy was employed, resistance collapsed within days. Yet the determining factor in the length of resistance had little to do with the strategic principles employed; rather, it had to do with the fact that the French continued to control a sufficient amount of territory to mount further resistance in 1870, while in 1940 they faced the imminent occupation of the entire nation. Fuller and Liddell Hart recognized that a nation's warmaking potential and not just its armed forces had to be eliminated before victory could be achieved when they advocated strategic bombing of population centers as a further means of weakening national resolve. But their belief that this could be achieved by only limited destruction once again reveals they vastly underestimated the resilience of modern nation states. Both men were undeniably correct in their assertions that total war was a major threat to Western civilization, and that the extent of this threat was not fully perceived by many statesmen and generals of their day. They were also correct, quite obviously, in their prediction that tank forces operating on the principles they advocated would have a devastating impact on traditional armies. Where they erred was in their belief that the replacement of the strategy of attrition with the strategy of armored war would by itself measurably assist the process of reducing the threat of total war.

It is also ironic to note that for two men who placed such high premiums on subtlety and maneuver in war, Fuller and Liddell Hart were remarkable brazen and direct in their attacks on the strategy of attrition. At no time did they attempt to conceal the fact that their objective was not simply the expansion of the Royal Tank Corps or mechanization of the cavalry, but the total destruction of that strategic concept. They often seemed to go out of their way in their efforts to demonstrate their belief that the Napoleonic model of war was a recent innovation in military history, that it was an aberration of the true principles of war, which had been established long ago, and that they were merely trying to correct these recent errors. They no doubt believed that this effort to establish historical legitimacy gave added weight to their arguments, but it also placed an unbridgeable gap between themselves and the army. If Fuller and Liddell Hart were correct, the army was guilty of gross mismanagement in its conduct of the World War I; if the army was correct, Fuller and Liddell Hart were merely ivory-tower theorists who had lost touch with the reality of modern war. There was no way these two positions could be reconciled.

The British generals of World War I and their successors of the interwar period were certainly as concerned about preserving their reputations as any other group of human beings in places of public authority and responsibility. But to see their opposition to Fuller and Liddell Hart's theories merely as attempts at self-justification is questionable. The concern at this point is not whether the generals were right or wrong, but whether they sincerely believed they were right. The weight of the evidence is that they did. These men had been raised and had spent their entire professional lives under the influence of the strategy of attrition, and, given the nature of military thought at this time, critical thinking was scarcely one of their stronger attributes. They had just emerged victorious—if such a term can be used—from the greatest military effort in their nation's history, and their confidence that they had acted according to the correct principles of war was only natural. In their view, the dominant fact of the war was that they had been pitted against the single most powerful army in Europe supported by all the forces of modern technology and industry. The result was "modern war" in all its horror. The casualties were a product of that fact alone, not the decisions of the generals or the principles on which they based those decisions. They believed they were no more responsible for the suffering and death of their soldiers than a physician is for the suffering and death of a patient with an incurable disease. This was the ultimate significance of Foch's statement that "modern war returns to the principle of a decision by arms and no other can be accepted." It removed the burden of responsibility from the shoulders of the generals and placed it on "modern war." Fuller and Liddell Hart's unpardonable sin was to claim that this responsibility should be placed on the generals. This the army could not, would not, and did not do.

This rejection of Fuller's and Liddell Hart's interpretation of the conduct of the World War formed the foundation for the army's contention that the strategic principles of the doctrine of armored war were totally unsound. Repeatedly, the army argued that by combining an intelligent use of terrain, antitank weapons, and infantry with the mobility of railroads and mechanized road transportation, it would be relatively easy to block the movement of an armored force operating behind the front lines. According to this line of reasoning, the armored force would then find itself in the unenviable position of having to choose between a long circumvention of the blocking units, consuming a tremendous and possibly fatal amount of fuel, or a direct encounter, negating the tanks' mobility and making them easy prey for artillery. Since Western Europe was both the most likely theater in which Britain would have to face tanks and also a heavily industrialized region with a highly developed transportation system, there was little doubt that a conventionally organized army with mechanized transport would be capable of dealing with such tactics.[99]

These arguments were not totally devoid of logic or reason, but they were based on the conviction that the only way to upset the equilibrium of the main body of an enemy army was by the action of the main body of one's own army. In the British Army's view, any attempt to employ the strategy of armored war ignored this basic truth and would immediately place a commander on the horns of an impossible dilemma. If he stripped his army's main body of sufficient offensive power for an armored thrust to succeed before the enemy had been weakened through attrition, he would risk the possibility of either exposing his own main postion to a decisive attack or having his armored force countered by massive reserves from the enemy's main body, which would be free to move as they wished. If, on the other hand, he did not weaken his main body, the armored force would lack the power to make a decisive thrust and could be countered easily by the enemy's local reserves. The weakness of this argument, as Fuller and Liddell Hart pointed out, was that the effect of a powerful armored thrust deep into an enemy's rear would be felt so rapidly that forces trained and equipped along traditional lines would be unable to react effectively. But the army refused to be swayed. It refused to accept the thesis that strategic paralysis could be produced by anything short of wearing out the enemy's reserves through a process of attrition, or that this could be accomplished quickly or cheaply when major powers were engaged.

This was the crucial dividing point between the British Army and the tank advocates throughout the interwar period. And it was this and not reverence for the horse or shortsighted government policy which determined the size, organization, equipment, and role of Great Britain's armored forces. Because the army adhered to the strategy of attrition, it logically concentrated its efforts on maximizing the combat power of those elements which would constitute the main body of any expeditionary force. This meant concentrating on the traditional combat arms, but it did not mean ignoring tanks. Rather, they were viewed as essential in supporting infantry attacks, and, in the 1930s, in replacing horsed cavalry for reconnaissance, screening, pursuits, and even raids. What the army did oppose was the creation of a powerful, independent tank arm, but the reason for this had nothing to do with a dislike of tanks as such; rather, it was simply due to their belief that such a force by itself could not play a decisive role in the strategy of attrition. In short, the tank controversy of the interwar years was much less concerned with the adoption of the new weapon than it was with the adoption of a new and radically different concept of strategy. The development of Great Britain's armored forces during the interwar years bears this out.

Notes

1. Biographical material is from J. F. C. Fuller, *Memoirs of an Unconventional Soldier* (London: Ivor Nicholson and Watson, Ltd., 1936) and Trythall, *"Boney" Fuller.*
2. J. F. C. Fuller, letters to his parents, 1894–98, file IV/1, items 1–16, Fuller Papers, King's College, University of London.
3. There are *Yoga: A Study of the Mystical Philosophy of the Brahmins and Buddhists* (London: W. Rider, 1925) and *The Secret Wisdom of the Quabalah* (London: Rider, 1937.)
4. Lieutenant General Sir Charles Broad, personal interview with the author at Beaminster, Dorset, August 19–20, 1970. General Broad was one of Fuller's classmates at the Staff College and witnessed this incident.
5. Anonymous, "Articles in Reviews," *Army Quarterly* 2 (October 1921): 177.
6. J. F. C. Fuller to B. H. Liddell Hart, 3 June 1924, Liddell Hart Papers.
7. See below, pp. 133–37.
8. J. F. C. Fuller, *The Reformation of War* (New York: E. P. Dutton and Co., 1923), p. 62.
9. J. F. C. Fuller, *The Dragon's Teeth: A Study of War and Peace* (London: Constable and Co., Ltd., 1932), pp. 28–36, 119–24.

10. J. F. C. Fuller, *War and Western Civilization: 1832–1932* (London: Duckworth, 1932), pp. 26–28, 181.

11. The essays were "The Application of Recent Developments in Mechanics and Other Scientific Knowledge to Preparation and Training for Future Warfare on Land," *Journal of the Royal United Service Institution* 65 (May 1920): 239–74; and "Future Naval Tactics," which was not published. See J. Luvaas, *The Education of an Army* (London: Cassell, 1965), p. 345.

12. Fuller to Liddell Hart, March 27, 1923, Liddell Hart Papers.

13. J. F. C. Fuller, *The Army in My Time* (London: Rich and Cowan, Ltd., 1935), p. 177.

14. Ernest Swinton to Liddell Hart, December 5, 1929; see also Hobart to Liddell Hart, October 7, 1934, Liddell Hart Papers.

15. Montgomery-Massingberd to Liddell Hart, 27 April 1926, Liddell Hart Papers.

16. Montgomery-Massingberd to Liddell Hart, 11 January 1927, Liddell Hart Papers.

17. Liddell Hart to Montgomery-Massingberd, 24 April 1926, Montgomery-Massingberd Papers, file 155, King's College, University of London.

18. Swinton to Liddell Hart, 5 December 1929, Liddell Hart Papers. Liddell Hart's letter cannot be found, but Swinton's tone indicates he was answering specific questions put to him. Trythall, *"Boney" Fuller,* pp. 158–59, 271, reaches the same conclusion.

19. Liddell Hart, *Memoirs,* 1:8.

20. B. H. Liddell Hart, "The 'Man-in-the-Dark' Theory of Infantry Tactics and the 'Expanding Torrent' System of Attack," *Journal of the Royal United Service Institution* 66 (February 1921): 1–22.

21. This refers to the interwar period only.

22. Edmund Ironside to R. Macleod, December 11, 1935 and January 5, 1936, Ironside Papers, File III, King's College, University of London.

23. B. H. Liddell Hart, "Suggestions on the Future Development of the Combat Unit: The Tank as a Weapon of Infantry," *Journal of the Royal United Service Institution* 64 (November 1919): 666.

24. Liddell Hart to Fuller, January 31, 1922, Liddell Hart Papers.

25. Fuller to Liddell Hart, August 6, 1925, Liddell Hart Papers. The reference is to Fuller, *The Foundations of the Science of War* (London: Hutchinson and Co., Ltd., 1926).

26. Liddell Hart to Fuller, 9 July 1929, Liddell Hart Papers.

27. Fuller to Liddell Hart, 12 July 1929, Liddell Hart Papers. Another disagreement, the nature of which is unknown, led Fuller completely to ignore his relationship with Liddell Hart in his memoirs. Despite the fact that they had corresponded for over a decade and Fuller was godfather to Liddell Hart's son when the memoirs were written, the latter's name never appears and there are only a few insignificant references to the military correspondent of the *Daily Telegraph,* who was, of course, Liddell Hart.

28. Ironside to Macleod, July 18, 1957, Ironside Papers, file III.

29. Lady Kathleen Liddell Hart, conversation with the author at States House, Medmenham, Buckinghamshire, August, 1970. The bitterness became so severe after World War II that whenever Hobart visited Lidell Hart's home, his first action was to go to the study, where photographs of the tank advocates were hung, and turn Martel's picture to the wall! He refused to allow it to be turned around as long as he was there.

30. Interview with Lieutenant General Sir Charles Broad at his home, Beaminster, Dorset, August 19–20, 1970.

31. Ibid. See also Liddell Hart, "An Evaluation of Senior Army Officers" (an unpublished and untitled memorandum prepared for the Secretary of State for War), 28 August 1937, Liddell Hart Papers, file A4/16.

32. General Sir Frederick Pile, personal interview with the author at the Cavalry Club, London, on 6 August 1971. See also Liddell Hart, "An Evaluation of Senior Army Officers," 28 August 1937.

33. Interviews with Generals Pile and Broad. On Hobart, see also Macksey, *Armoured Crusader.*

34. Interview with General Pile.

35. Liddell Hart, *Paris; or the Future of War* (London: Kegan Paul, Trench, Trubner and Co., 1925), p. 10; Fuller, *The Army in My Time,* p. 164.

36. Liddell Hart, *The Ghost of Napoleon* (New Haven, Conn.: Yale University Press, 1935), pp. 105 and 123. The theme of the entire book is that prewar military thought was based on false principles.

37. Liddell Hart, *The Remaking of Modern Armies* (Boston: Little, Brown and Co., 1928), p. 101.

38. Liddell Hart, *The British Way in Warfare* (London: Faber and Faber, Ltd., 1932), p. 18. For a similar assessment by Fuller, see *Armored Warfare* (Harrisburg, Pa.: The Military Service Publishing Co., 1943), pp. 44–45. This is a reprint of *Lectures on F. S. R. III* (London: Sifton Praed, 1932).

39. Liddell Hart, *The Ghost of Napoleon,* pp. 120, 128–29.

40. Liddell Hart, *Through the Fog of War* (New York: Random House, 1938), p. 74.

41. Fuller, *Memoirs of an Unconventional Soldier,* pp. 28–29.

42. Ibid., pp. 143–44.

43. Liddell Hart, *Paris; or the Future of War,* p. 52.

44. Liddell Hart, *The Remaking of Modern Armies,* p. 17.

45. Fuller, *On Future War* (London: Sifton Praed and Co., Ltd., 1928), pp. 168–69.

46. Liddell Hart, *Paris; or the Future of War,* pp. 34–35; Fuller, *The Reformation of War,* pp. 107–8.

47. Fuller, *The Reformation of War,* p. 148.

48. Liddell Hart, *Paris; or the Future of War,* pp. 24–25, Fuller, *The Reformation of War,* pp. 28–30.

49. Liddell Hart, *A Greater Than Napoleon: Scipio Africanus* (London: William Blackwood and Sons, Ltd., 1927), pp. viii, 192–96.

50. Fuller, *The Reformation of War,* pp. 14–19; *The Dragon's Teeth* pp. 167–72, 293–97.

51. Fuller, *The Dragon's Teeth,* pp. 158–207.

52. Fuller, *The Generalship of Ulysses S. Grant* (New York: Dodd, Mead and Co., 1929), pp. 397–400.

53. Fuller, *The Dragon's Teeth,* pp. 88–89, 107–9.

54. This discussion first appeared in Liddell Hart, *The Decisive Wars of History: A Study in Strategy* (London: G. Bell and Sons, Ltd., 1929), pp. 223–62. It was expanded and refined in *The British Way in Warfare,* pp. 73–99.

55. Liddell Hart, *The British Way in Warfare,* p. 98.

56. Ibid.

57. Ibid., p. 99.

58. Ibid., pp. 99–114.

59. Liddell Hart, *The Decisive Wars of History.* The book has gone through a number of editions, some under different titles, but without any change in Liddell Hart's thesis. All references here are to the latest edition, *Strategy; The Indirect Approach* (London: Faber and Faber, Ltd., 1967).

60. Ibid., p. 161.

61. Ibid., pp. 145–46.

62. Fuller, *The Reformation of War,* pp. 109–11, 154–59; Liddell Hart, *Paris; or the Future of War,* pp. 50, 76; *The Remaking of Modern Armies,* pp. 123–24.

63. Fuller, *The Reformation of War,* pp. 243–45; Liddell Hart, *The Remaking of Modern Armies,* pp. 15–17.

64. Fuller, *On Future Warfare,* p. 229. See also *Armored Warfare,* pp. 2–5.

65. The most important parts of Plan 1919 were printed in Fuller's *Memoirs of an Unconventional Soldier,* pp. 321–32.

66. Fuller, *The Reformation of War,* p. 118; Fuller Papers, Special Collections Department, Rutgers University Library, New Brunswick, file 1 contains Wilson's forwarding letter of July 20 and a copy of Foch's approval dated 6 August 1918.

67. Fuller, *On Future Warfare,* pp. 83–105. See also Liddell Hart, *Paris; or the Future of War,* p. 79.

68. Fuller, *Tanks in the Great War* p. 312.

69. G. Martel, *Our Armoured Forces,* pp. 396–402.

70. Fuller, "The Application of Recent Developments in Mechanics," pp. 261–62.

71. *Daily Telegraph,* March 31, 1926.

72. Liddell Hart, *The Remaking of Modern Armies,* pp. 61–71.

73. Liddell Hart, *Paris; or the Future of War,* p. 89. He specifically criticized Fuller for his failure to see the need for infantry support in the offense in "Notes on Fuller's 'Lectures of F. S. R. III,'" an unpublished and undated memorandum apparently written at the time that his book was published in 1932. Liddell Hart Papers.

74. Fuller, *The Reformation of War,* pp. 160–62; *Towards Armageddon: The Defense Problem and Its Solution* (London: Lovat Dickson, Ltd., 1937), pp. 138–39.
75. Sir John Moore was a British officer who served in the American Revolution and Napoleonic Wars. In 1803, he formed the Light Brigade, a select unit armed with rifles and trained to act as light infantry. The unit, enlarged and designated the Light Division, went on to win fame as one of the finest units in the British Army. Moore was killed at the battle of Corunna in 1808.
76. Liddell Hart, *The British Way in Warfare,* pp. 131–32; *The Future of Infantry* (London: Faber and Faber, Ltd., 1933), pp. 35–49; Fuller, *On Future Warfare,* pp. 169–70.
77. Liddell Hart, *The Remaking of Modern Armies,* p. 60.
78. Fuller, *Lectures on F. S. R. III* (London, Sifton Praed Co., 1931). It was later published as *Armored Warfare* (Harrisburg, Pa.: The Military Service Publishing Co., 1943), from which all references are taken.
79. Ibid., pp. 17–24.
80. Ibid., pp. 32–41.
81. Ibid., pp. 26–27.
82. The same views are expressed in Liddell Hart, *The British Way in Warfare,* pp. 199–200.
83. Fuller, *The Reformation of War,* p. 240. Fuller estimated the process would take twenty years because of military conservatism.
84. Fuller, *On Future Warfare,* pp. 371–89.
85. Fuller, *Memoirs,* p. 475.
86. Fuller, *Generalship: Its Diseases and Their Cure* (Harrisburg, Pa.: Military Service Publishing Co., 1936), p. 28.
87. Fuller, *The Army in My Time,* pp. 160–62.
88. Fuller, *Memoirs,* p. 151.
89. Liddell Hart, *The Ghost of Napoleon,* p. 145. See also Fuller, *Generalship: Its Diseases and Their Cure,* pp. 31–35.
90. Liddell Hart, *The Ghost of Napoleon,* pp. 177–85; *The Remaking of Modern Armies,* pp. 76–78.
91. "Bardel," "Study and Reflection vs. Practical Experience," *Army Quarterly* 6 (July 1923): 318–31. Fuller's criteria are found in *Generalship: Its Diseases and Their Cure,* pp. 31–35.
92. Liddell Hart, *Sherman* (New York: Frederick A. Praeger, 1960), pp. 6, 7. The original edition was published in 1929.
93. Fuller, *Generalship: Its Diseases and Their Cure,* pp. 97–106.
94. Liddell Hart, *The Remaking of Modern Armies,* pp. 181–91.
95. Fuller, *Generalship: Its Diseases and Their Cure,* p. 97–106.
96. Liddell Hart, *Europe in Arms* (New York: Random House, 1937), pp. 181–82.
97. Heinz Guderian, *Panzer Leader* (New York: E. P. Dutton and Co., 1952), p. 20; Colonel Dr. J. L. Wallack, Comments on "Voice of Israel" broadcast, 15 July 1967. I am indebted to Dr. Charles Hall, U. S. Army War College, Carlisle Barracks, Pa., for a transcript of this broadcast.
98. During the battle of Cambrai, a group of advancing British tanks became separated from their infantry near the village of Flesquieres and twelve were knocked out by a German battery manned by stragglers. Haig, visiting the site a few days later, noted "the importance of Infantry operating with Tanks" to protect them from such fire, a principle as true today as it was then. Douglas Haig, *The Private Papers of Douglas Haig,* ed. Robert Blake (London: Eyre and Spotteswoode, 1952), p. 269.
99. For a more detailed account of these arguments, see below, pp. 116–17.

• 4 •

The Postwar Period: 1918–27

Introduction

THE future course of mechanization in general and of tank development in particular captured the army's spotlight almost immediately following the end of the war. Within a few months of the armistice, the military journals began publishing articles claiming that tanks had won the war, and by mid-1919 rebuttals were also appearing. The striking feature of this controversy was its public nature. The *Journal of the Royal United Service Institution,* the *Army Quarterly,* the *Journal of the Royal Artillery,* the *Royal Engineers Journal* and even the *Cavalry Journal* published articles reflecting every point of view. More significantly, several of these journals sponsored essay contests inviting officers to explore various aspects of the mechanization question. Since many of those who won or placed in these contests argued for mechanization, it is difficult to argue that the advocates of the new arm did not get a fair hearing.

Yet the theory of armored warfare emerged only slowly from this debate. Although the germ was contained in Fuller's Plan 1919, he gave it only minor attention in his immediate postwar writings. In fact, it did not emerge fully until the publication of his *The Reformation of War* in 1923 and Liddell Hart's *Paris, or the Future of War* two years later. In the meantime, the army struggled with the myriad of problems brought about by demobilization and the introduction of tanks into the army. Yet, while progress was made in many areas of tank development and Great Britain continued to be the world's leading tank power, the British Army could still only muster one battalion of tanks by 1927.

The Postwar Controversy

The confusion and uncertainty over the tank's future was to some degree natural, given the circumstances of its birth. Because it was a product of the Western Front, there was a widespread assumption in the army that its future development depended upon the likelihood of a repetition of that experience. Moreover, the tank was not only a product of the Western Front; it was also conditioned by it. Because its role had been limited to clearing a path through the enemy's trench system for the infantry, the army had considered its speed, radius of action, and durability to be inconsequential. As Ellis recalled shortly after the war, designers had been told their tanks should "run fifty miles and might then fall to pieces."[1] This sufficed at Cambrai and Amiens; it did not in 1919, when the army was studying the tank's future.

Fuller's contribution to this early debate did not always enhance the clarity of the theory of armored warfare, for, in the articles he wrote for professional journals at this time, he minimized the changes he sought to make in the army's strategic and tactical doctrines. In his Gold Medal Prize essay for the *Journal of the Royal United Service Institution* his theme was that the army should replace muscle power with machine power wherever possible to increase its efficiency. Arguing that "tools or weapons, if only the right ones can be invented, form 99 percent of victory," he asserted that tanks could replace the infantry and cavalry, reduce the need for field engineers, supplement the artillery, and magnify the power of machine guns; in short,he argued, tanks should become the focus around which the rest of the army would be organized. They would give the army far greater combat power with less manpower and would enable it to easily outclass any opponent organized and equipped along traditional lines.[2] Radical as these ideas were—and many of Fuller's contemporaries reacted with outraged disbelief at his proposals[3]—the fact is that the strategy of armored war received relatively short shrift in the context of the article as a whole. He briefly described its basic principles as part of a general commentary on the evolution of wartime tactics, noting only that "stupendous revolutions in tactics were actually in progress when the Great War ended" (a reference to Plan 1919); but he did not place it at the center of his arguments nor did he use it to justify the expansion of the army's tank forces.[4]

Similarly, in a series of articles published in the *Cavalry Journal* in 1920,[5] Fuller proposed to replace horsed cavalry with tank forces that merely would perform the same functions. Indeed, in this instance he declared the timeless validity of the strategy of attrition! Fuller's thesis in these articles was quite simple: throughout history, the objective of tactics had been "the giving of blows without receiving them" and that success had gone to those commanders who were able to combine the elements of stability and mobility to achieve this end; that is, those who were able to pin down the enemy and annihilate his forces with a decisive blow struck by a powerful mobile force.[6] Labeling this concept the "Cavalry spirit," Fuller traced its evolution in a wide-ranging historical analysis from ancient Greece to the World War. The principle, he argued, was essentially correct, but it had become impossible to implement because increased firepower of infantry and the absence of flanks had combined to destroy the mobility of the horse. Fuller's solution, naturally, was to replace horsed cavalry with tanks, but he maintained that this would not destroy the cavalry , but, rather, restore the "Cavalry spirit" to its rightful place as the decisive act in war:

> The battle remains the same—there is the assembly, the approach, and the attack. The fight remains the same—there is the act of demoralization, the act of decision, and the act of annihilation. Grand tactics remains the same—there is attrition, surprise, envelopment and penetration.[7]

As a whole, these articles represent one of Fuller's rare attempts to convert others to his point of view by showing some concern for their feelings. Although it is doubtful that he met with any success, it is noteworthy that, along with the disclaimer that the articles did not reflect the views of the general staff, the editor of the *Cavalry Journal* described Fuller as "a deep student of military history and a rare thinker." He continued with the comment that Fuller had "contributed largely to the success of the Tank Corps in the recent war by his clear vision of the possibilities of the youngest branch of the Service."[8]

In his books however, Fuller was less constrained. In *Tanks in the Great War,* published in 1920, he provided the graphic description of armored war that has already been noted. It was not until three years later, however, in *The Reformation of War,* that the concept was placed fully in its sociopolitical setting. With the publication of Liddell Hart's *Paris; or the Future of War* in 1925, which was a direct assault on the strategy of attrition, the essentials of the strategy of armored war were complete.

In the meantime, the controversy over the future of tanks continued to rage in British military journals. Fuller's Gold Medal essay was seconded by another essay that was also strongly protank. Authored by a Captain W. D. Croft, it was more practical in its approach but no less committed in its outlook. Croft closely paralleled Fuller's arguments regarding the advantages of tanks over horses, but he gave no indication that he perceived tanks could be used to attack the enemy's brain. Indeed, he declared that the tank's chief advantage was "the speed of its advance from its area of concentration to its objective, the enemy's field armies."[9]

The *Journal of the Royal United Service Institution* was not alone in sponsoring studies of the future possibilities of mechanization. In 1920, the *Journal of the Royal Artillery* invited entries on the subject of "The Reorganization of the Royal Regiment of Artillery with a View of Obtaining in War the Best Value from Modern Equipments with a Minimum Expenditure of Human Energy." Although the question did not specifically refer to mechanization in general or tanks in particular, these matters were inherent in the subject, and most of the essays meriting publication dealt with them. The respondents, however, reflected the confusion of the time more than they offered any likely avenues of development. Of the five essays published, the first three were conventional in their approach: none opposed mechanization, but all were convinced that the horse still had a role in war and concentrated on technical questions such as the standardization of artillery equipment.[10] The fourth and fifth essays, however, were strongly protank. Both writers expressed the conviction that the days of the horse were numbered and that they would be replaced by tractors. Both also adovcated placing light cannon on tractors, reasoning that only such vehicles would have the mobility to cope with the threat of fast-moving enemy tank forces.[11]

There was one area, however, in which many officers from all branches of the army were coming to see the need for mechanization: the supply and administrative services. This, of course, would not mean the supercession of the traditional arms, but the enhancement of their strength; and many who were skeptical about the value of the tank itself felt that mechanization should definitely proceed as rapidly as possible within these latter areas. This viewpoint was reflected in the comments made following a lecture given at the Royal United Service Institution by Lieutenant Colonel Phillip Johnson early in 1921.[12] An outspoken advocate of mechanization and the designer of the Medium D tank on which Fuller had based his Plan 1919, Johnson made the best possible case for the new arm without winning many converts among his audience. In the discussion period following his presentation, skeptics challenged many of his assumptions, but they expressed common agreement on the value of mechanization of the support services. Their opinions were summarized by Major General A. L. Lynden Bell, the chairman for the lecture, who, although he stated that no tank could ever "replace flesh and blood and the horse" in combat, strongly supported the latter reform as it would enable the army to move supplies far easier and faster than ever before and would thus be a great step forward.[13]

Another possibility that began to receive considerable attention at this time was that of tank-cavalry cooperation in combined operations. This idea was first broached in 1921 by a writer in the *Journal of the Royal United Service Institution.* He asserted that the armor and firepower of tanks would enable them to protect the cavalry when they came under fire, while the mobility, radius of action, and capacity for close reconnaissance of the cavalry would enable them to provide added security for the tanks.[14] A more detailed study of such cooperation was made the following year in an article that described a theoretical operation of both arms. In this article, the author balanced the ability of tanks to move faster and over greater distances in good terrain with the ability of cavalry to penetrate rough terrain where the tanks could not go, and the vulnerability of cavalry to hostile fire with the inability of tanks to provide their own protection in rugged country. The obvious solution, he concluded, was for both arms to work together to make up for the deficiencies of the other. The tanks would provide additional firepower for the cavalry and could attack entrenched enemy positions when necessary, and the latter could carry out close reconnaissance in rough country and provide security for the tanks.[15]

Both of these ideas—mechanization of the support echelon and combined tank-cavalry operations—were to play an important role in the development of Great Britain's armored forces during the interwar years. Although the first did not deal directly with the tank forces, it was obviously a necessary corollary to them even in the limited form envisioned; tanks are of little value if gas must be brought to them in horse-drawn wagons. Even though there was strong disagreement over the ultimate form of the army, the tank advocates and more conventional officers found themselves in agreement on this point and could work together.[16] On the second question, combined tank-cavalry operation, agreement proved more difficult. Wartime experience had demonstrated that in open country devoid of enemy troops, cavalry could easily outpace tanks, but that enemy fire could quickly immobilize cavalry and prevent them from keeping pace with the tanks. No solution to this problem was ever devised, but the idea continued to find supporters in many circles within the army. Its most important effect, however, was to convince many tank advocates that cooperation with other arms was detrimental to the development of their own arm and to push them further toward the all-tank concept.

The cavalry, of course, did not lack for defenders, as we have already seen.[17] Fuller's articles in the *Cavalry Journal* brought forth a number of responses attacking his claim that the day of the horse was over. The most thoughtful of these, by Brigadier General B. A. Weir, focused on the assumption that the experience of the Western Front should be treated as typical of modern war. He pointed to the Eastern Front and Palestine theater during the World War and the civil war then being waged in Russia as examples of fighting in which a single, continuous front did not exist. His conclusion was that "we have to train our cavalry for a campaign which is much more likely to resemble in its broad outline that in Palestine than that in France."[18] The cavalry also received support from within the House of Commons. In a 1921 debate on the aboliton of four cavalry regiments, an impassioned cavalryman by the name of Sir Keith Fraser defended retention of the units on grounds that would have made any reduction impossible. He argued that, of the four regiments, one was Irish and it would inflame opinion there to abolish it; another was commanded by Sir John French and it would insult him to abolish that one; and that the two remaining regiments had served in India for many years while other units had enjoyed easier service at home and it would insult the memories of those who had served abroad for so many years to abolish them.[19]

It seems, however, that the defenders of the tank gradually gained ground among thinking soldiers. For its 1924 military prize essay, the *Army Quarterly* asked participants to write on the possible composition of a future expeditionary force, and, although the question of tanks was not raised explicitly, it figured prominently in many of the responses. Of the fourteen entries, nine held that the tank and not the infantry would be the "predominant arm" of the future, and two even recommended that the infantry be totally eliminated. In addition, seven, including the winning essay, expressed the conviction that the day of the cavalry was over and that the horse should be replaced by the tank.[20] The winning essay was particularly interesting. Appearing in the July 1923 issue, it advocated the creation of infantry divisions composed of four tank and twelve infantry battalions carried in armored tracked vehicles. Although the size of such a division would make it extremely difficult to command effectively, this organizational concept was certainly highly progressive for the time.[21] The fact that it could win the competition also indicates that the *Army Quarterly* and the army in general were not so rigid as often assumed.

The most reliable source for the army's attitude toward tanks at this time, the postwar *Field Service Regulations,* also reveals that the army had accepted tanks as a valuable instrument of war. Emphasizing their firepower and mobility, the *Regulations* described tanks as "essentially a weapon of offense." It admitted—accurately—that their size, noise, and inability to negotiate rough terrain were serious weaknesses, but it treated these as matters that could be resolved by a variety of tactical solutions. It placed far greater emphasis on the advantages the weapon provided. Specifically, it stated that tanks were capable "of delivering a large volume of accurate fire during movement," of moving "over country where roads and tracks do not exist," of "cross[ing] trenches and surmount[ing] obstacles," of "moving through entanglements . . . [and crushing] down the wire to form lanes passable by infantry in single file," and of destroying "hostile weapons and personnel by passing over them." It concluded that "its moral efects on hostile troops is very great" and the "the power of delivering successful surprise attacks against almost any type of defenses is one of the most important advantages of use of tanks in large number."[22]

Turning to a description of tank operations, the *Field Service Regulations* stated that the duties of tanks in the attack were: "(i) To assist the advance of infantry. (ii) To destroy hostile tanks. (iii) To exploit a success." To perform their duties successfully, it stressed that the cooperation of tanks with the other combat arms was essential. Artillery support was necessary to protect tanks from enemy guns by means of counterbattery fire and the laying of smoke screens. But it placed greater emphasis on the closest cooperation with infantry on the grounds that "tanks must protect infantry from machine gun fire and the delay imposed by uncut wire; infantry must protect tanks from the close range fire of enemy field artillery and antitank guns." In addition, it specified that attacking tanks should be allotted specific objectives within the zone of attack, that the organizational integrity of their units never should be disrupted, and that in continuous operations, a system of successive echelons should be established to allow units to regroup after each assault. Defensively, the role of tanks was admittedly limited. They were described as being "chiefly valuable in keeping alive the offensive spirit at times when the defense is weakening" and were assigned the tasks of conducting limited counterattacks, engaging enemy tanks, and covering the withdrawal of infantry when this became necessary.[23]

The *Regulations* also gave a prominent role to the cavalry, however. It was described as having the "ability to move rapidly, to cover long distances in a comparatively short time, and to move across country."[24] Offensively, its duties were reconnaissance before the opposing forces came in close contact and particularly in rough or heavily wooded terrain; attacks or threats against the enemy's flanks or rear; and the maintenance of pressure on the enemy once he began to retreat and during the pursuit. Defensively, their duties were to form a mobile reserve in support of the main positions and to act as a rearguard or delaying force if and when retreat became necessary.[25]

Although the *Regulations* did not attempt a comparison of tanks and cavalry, it is clear that the postwar army regarded the two arms as complementary rather than competitive. Both relied on the combination of firepower and mobility to accomplish their missions, but not in the same proportion; tanks had the greater firepower and protection, while cavalry had the greater mobility. As a result, the tasks assigned to tanks were those of a main battle weapon, while those given to the cavalry were of a peripheral but equally vital nature. The analogy may be made—and this is inherent in what the *Regulations* stated rather than explicitly declared—with the roles of heavy and light cavalry during the Napoleonic era. Tanks were like the heavy cavalry, which, in conjunction with infantry and artillery, engaged the enemy head-on in an attempt to smash his main positions; the cavalry, on the other hand, was comparable to the old light cavalry, which harassed the enemy and provided information, but avoided any major confrontations against his strength. As long as the cavalry could successfully assert that it could fill this role, its position, though limited, remained secure.

Most importantly, however, the postwar *Field Service Regulations,* whether describing the role of tanks or cavalry, unhestitatingly reaffirmed the army's belief in the Napoleonic model of war and the strategy of attrition. As described in the 1920 and 1924 *Regulations:*

1. Battle is the decisive factor in war. [It is not until the enemy's power of resistance has been finally crushed in battle that the objects for which a country has entered into war can be secured.]
2. Operations in every campaign will usually follow the following distinct phases:
   (i) The advance to the battlefield, including the first contact between the opposing advanced troops.
   (ii) The [deliberate] battle between the main forces, which, in the

case of national armies, may extend over [an indefinite] (a lengthy) period.

(iii) The withdrawal and pursuit, the extent of which will be in direct proportion to the success of the victor [in the deliberate battle]. These phases alternating with occasional periods of comparative inactivity on the whole or portions of the front my occur again and again during the course of a compaign until the power of one side or the other is broken and decisive victory is obtained.[26]

In sum, then, it is clear that, while the British Army had fully accepted tanks as a valuable weapon of war by the mid-1920s, it also continued to have faith both in horsed cavalry and in its traditional strategic and tactical doctrines. The formulation of the strategy of armored war as a fully developed strategic theory wrought little change in this state of affairs. The appearance of the first volumes of the army's account of the war provided the historical justification for this refusal to change,[27] and in 1927 there appeared the most effective case made against Fuller's and Liddell Hart's writings to date. Entitled *The Mechanization of War,* it was written by Victor Germains, a popular writer on military affairs who, even when he was patently wrong, presented the image of a well-informed and sensibly progressive student of war. Germains was unique because the range of his criticism ran the gamut of the tank advocates' theories. Despite this, however, he never denied the value of tanks as such; rather, his aim was to correct what he asserted were the greatly exaggerated claims of the tank advocates by placing the weapon's strengths and weaknesses into a proper balance. Regarding Fuller's and Liddell Hart's charge that the mechanization of civil life necessitated a similar change in the army, for example, he countered that this had been going on for many years and had had and continued to have the army's full support. The tank advocates were confused, he explained, because they failed to see that mechanization meant the adoption of machine power in any form, and their claim that the army opposed such changes simply because it refused to abolish everything except tanks was ridiculous. Furthermore, Germains claimed that the application of this argument to cross-country tracked vehicles was fallacious because such machines were not an integral part of modern industrial civilization. Their application to civil life was severely limited, and if they were developed, it would have to be solely for war.[28]

In general, however, Germains concentrated on the tanks's technical problems, most of which he believed were insurmountable. This concern was natural. Slow, cumbersome, and unreliable, the first tanks were a terribly fustrating weapon with which to work. Even one of the leading tank advocates later admitted that there was much truth in the wartime Fifth Army complaint that "it took 12,000 pounds to make a tank and 2,000 pounds of twelve inch howitzer ammunition to destroy it after the Germans had turned it into a pill box."[29] Germains charged that the tank advocates' claim that many of the weapons difficulties could be overcome by advances in technology was unrealistic. He specifically called attention to the problems of tank design, the power of antitank weapons, the restrictions of terrain, and the difficulties of supply. With regard to design, he asserted that it was impossible to reconcile the elements of armor, armament, speed, and endurance because any improvements in one would lead to greater difficulties in one or more of the others. An increase in protective armor, for example, would decrease the tank's speed, so that its overall vulnerability would remain relatively the same. In any event, he was certain that antitank guns would dominate tanks on future battlefields simply because no tank could hope to outrun a high-velocity projectile. Because only so much weight could be added to a tank and because the entire shell had to be protected and not just the "vitals" as in a warship, he concluded that it was impossible to make tanks immune to the fire of antitank guns. In one passage, Germains reasoned that a heavy machine gun could penetrate a heavy tank at 500 yards and, since it would take the tank one minute to cover this distance and the rate of fire of the machine gun was 300 rounds per minute, the tank would have to avoid 300 rounds, any of which could knock it out. Furthermore, Germains continued, the fire of the tank would not be nearly so effective as that of the antitank gun, which would be stationary and, being smaller, could be easily concealed and thus more difficult to locate. The problem would be vastly increased by the addition of many antitank guns and by improvements in pillbox design then being explored. As to the suggestion that tanks be armed with their own heavy guns to counter antitank weapons, Germains derisively commented that "an elephant hunter does not climb on the back of another elephant." Such guns would be far more effective if they were left on the ground, where they would retain the advantage of concealment.[30]

In the matter of the mobility of armored forces, Germains again challenged every one of the tank advocates' claims. First, he questioned their emphasis on the importance of mobility itself, noting that, while it was not a insignificant factor in war, it was ridiculous to assert that "superiority in mobility *in itself* spells superiority in fighting power." Second, he denied that tanks would be able to move freely across country as ships moved the sea. He pointed to heavily wooded areas, swamps, and mountaineous regions as examples of locales where the movement of tanks would be, if not totally impossible, certainly very restricted. The placement of antitank guns in such areas, moreover, would make them "tank proof" and enable them to serve as bases from which to attack armored forces moving in the open. And, third, he challenged the assumption that armored forces would necessarily be more mobile than nonarmored forces. Assuming Great Britain faced a highly industrialized opponent as in World War I, he argued that the enemy would be able to make effective use of a complex rail and road network that could be used to counter any armored strike. He concluded: "By the skillful [*sic*] use of his railway net and road net, and given vigilance against surprise, there can be no doubt but that a well-equipped defender can always anticipate attack by a *relatively slow* cross-country mechanized force." Finally, Germains maintained that the tank advocates had ignored the question of logistics. He asserted that mechanized forces would be unable to live off the land as the cavalry could, and that they would actually require more supplies in order to keep them going. Moreover, such support could only come from a sizable "tail," which would inevitably become a drag and greatly hinder the mechanized force's mobility.[31]

Severe as these criticisms were, Germains nonetheless recognized that tanks could play a valuable role in future campaigns if properly coordinated with the traditional combat arms. He saw no such redeeming value in the theory of armored warfare, however. His arguments closely paralleled those of the army. They were based on the conviction that "Clausewitz's principle, that you can achieve victory in a war only by grappling with and decisively beating the enemy's main armed forces, is as true today as when it was written." He maintained that the great suffering and loss of major wars could not be avoided, and he derided those who held to the "popular fallacy . . . that you can gain victory in war by maneuver and not by fighting." As for the power of mechanized forces, he claimed that "to send a 'mechanized' column careening wildly about the enemy's country" would probably result in its being cut off, surrounded, and annihilated. An alternate scenario, he continued, might be for the enemy to counter an armored strike by attacking its base and forcing its return to protect its supplies and communications, much as Scipio Africannus forced Hannibal back to North Africa by attacking Carthage during the Second Punic War.[32] In any event, he wrote:

> it is difficult, then, to see how "mechanized" columns on land are going to produce a *decision*. What they gain in speed they lose in smashing power and in staying power. The speed of any army on land, like that of a fleet at sea, is that of its slowest, not its fastest unit. To throw our fast units far ahead, beyond supporting distance of the army as a whole, is to run the danger that these will come up against superior enemy forces, and be beaten before we can come effectively to their support.[33]

Germain's proposals for the future organization of the British Army followed logically from his analysis. He proposed that mechanization in its broadest sense continue throughout the army, but that the ratio of tanks to infantry remain at one tank battalion for each infantry division. He opposed any reduction in the strength of the infantry or supporting arms on the grounds that they were necessary to provide the army with the staying and smashing power it would require in combat. He also agreed, however, that a separate mechanized force might be a distinct advantage in certain operations such as raids, but he contended that it could not bring about a decision by itself and should only be an adjunct to the main body. In conclusion, he recommended that the army follow three broad lines of development: first, the maintainence of a thoroughly trained and efficient general staff; second, the establishment of an expeditionary force of twelve divisions capable of deployment on the Continent; and third, the creation of a backup force of an additional twelve divisions capable of supporting the initial force as soon as possible. All twenty-four divisions should be supported by tanks in the same proportion as noted above, and the replacement of horse transport by motorized transport should continue to its logical conclusion. To replace such an army with a small, wholly mechanized force, however, would be clearly suicidal.[34]

The impact of Germains's book is difficult to measure. Certainly, officers hostile to Fuller and Liddell Hart found comfort in what he wrote, but it is clear that many found his case less than completely satisfactory. The *Army Quarterly,* for example, told its readers that Germains had made the best possible case against mechanization, but that because it still had serious gaps, the weaknesses of that case were only more clearly highlighted. Indeed, it concluded that "the general impression made by the book is that Mr. Germains' knowledge of war and of military matters is theoretical rather than practical."[35] Although the journal did not point out any of these weaknesses specifically, one stands out in light of the army's experiences with tanks until that time: Germains's emphasis on the tank's vulnerability to antitank guns. Although the army certainly placed considerable faith in such guns, to prove their power by pitting a single gun against a single tank with no other factors influencing the outcome was unrealistic and, in fact, violated the army's tactical doctrines, which held that tanks should be employed in the largest numbers available and in conjunction with other arms. Given such lapses, the *Army Quarterly*'s reaction was hardly surprising.

The First Years of the Royal Tank Corps

It was one thing for the army to recognize the importance of tanks and to establish a doctrine to govern their operations; it was another and far more difficult task to decide how they should be organized and equipped and how the army could obtain a sufficient quantity of them to meet its needs. The problem was compounded by the confusion into which the army was thrown by the rapid postwar demobilization. The sudden and unexpected end of the war placed a great strain on the army's resources and afforded it little time to plan for the transition to a peacetime status. Indeed, the confusion was so great that some expected conscription to continue into the postwar period.[36] Such an asumption was totally unrealistic, however, and only added to the army's problems by forestalling the preparation of more realistic plans for demobilization.

During the war, Great Britain had raised and equipped a total of twenty-two tank battalions of all types, of which eighteen reached France before the Armistice. In addition, plans had been laid during the summer of 1918 for further expansion of the corps, and an order was placed for 6,000 tanks to support the 1919 campaign that all but the most optimistic then thought would be necessary.[37] The Armistice brought an immediate halt to these plans and marked the beginning of a precipitate decline in the strength and fortunes of the Tank Corps.

This decline was not, however, the result of antitank feeling at the War Office, but to Treasury restrictions, which the army unsuccessfully sought to overcome. On November 13, 1918, the Ministry of Munitions canceled all orders for the production of tanks. Two days later, Fuller, then deputy director of staff duties (tanks), formally protested this decision in a memorandum to his immediate superior, the DSD. Fuller argued that an immediate resumption of tank production was necessary because it was impossible to be certain that the Germans would live up to the terms of the Armistice. Working on the assumption that the postwar army would be maintained at a strength of twenty infantry divisions and that each division should have a tank battalion for support, he concluded the army's tank requirements should set at twenty battalions with a total strength of 500 machines. The DSD endorsed this proposal, adding that tanks would be useful in controlling civilian disturbances as well as in conventional military operations. He then forwarded it to the Army Council, which, on December 2, added its approval to the plan, although stipulating that half of the tanks in the program would come from wartime models then in production rather than all-new models as Fuller had proposed. Exactly what happened at this point is not clear, but the end result was that the army failed to receive any of the tanks it requested. It seems highly probable, however, that the funds were refused by the Treasury. On November 28 the assistant financial secretary at the War office informed the deputy CIGS, Major General Sir C. H. Harrington, that the Treasury would have to approve all requests for equipment before the orders were placed and the army could no longer operate on credit and expect automatic approval as it had during the war. The secretary also expressed the opinion that the cost of Fuller's program—some 2.5 million pounds—seemed extravagant in light of the fact that the army was completing development of a newer model of tank. The Army Council's decision to order only half the number of new tanks and complete the quota with models already in production was apparently an attempt to reduce the total cost of the package, but it had little effect on the Treasury's opposition.[38]

The consequences of this action were revealed in a memorandum on the state of the Tank Corps written by Fuller the following March. From its peak wartime strength of twenty-six battalions only six months before, the corps was now reduced to eight, five in France (which were equal to only three at wartime strength) and three in England (none of which was fit for operation). Moreover, the instructional staff at the tank training center had been reduced to the point where it was "practically inoperative"; the tank-repair facilities in England and France were being phased out and would cease to exist by June 1; and the corps was having considerable difficulty retaining personnel. Only 500 men thus far had reenlisted, and Fuller expected that the corps would be able to retain about 400 more at best. He hardly exaggerated when he concluded that "the Tank Corps will virtually cease to exist, except on paper, when general demobilization takes place."[39] Fuller's assessment was endorsed a few days later when Harrington wrote the secretary of state for war that "for the moment the life has gone out of both the personnel and the production side of the Tank Corps." He concluded: "It looks as if we shall want these tanks in different parts of the world and possibly at home and I should like your authority to put the Tank Corps on to an efficient basis again."[40]

Despite Harrington's plea, no new proposal for the reequipment of the Tank Corps came before the Army Council for consideration until October 1920. At that time, the MGO reported that the army required four types of armored tracked vehicles: a light infantry tank, a fast-moving cavalry tank with an extended radius of action, a heavy tank to direct infantry support, and a transporter to carry an eighteen-pound gun. The development of the first two types was progressing satisfactorily, and he hoped that in the case of the second a prototype of the Medium D could be completed by the beginning of the following year so that it could go into full production by April. He then placed two proposals before the council, the first calling for sufficient tanks to equip six battalions by May 1922 at cost of 1,300,000 pounds, the second calling for sufficient tanks for three battalions at half the cost. In the discussion that followed, the first proposal was supported by the DSD on behalf of the CIGS on the grounds that there was a critical shortage of armored vehicles in the army and that this was the only way to correct the situation. The other members of the council disagreed, however, arguing that such a program would mean deep cuts in the production of armored cars, which were also needed. At that point, the financial secretary intervened, saying that funds simply were not available to produce the number of tanks required for six battalions. The question was quickly solved, and the second program, calling for the production of tanks for three battalions, was adopted.[41]

Meanwhile, the army was trying to decide how to bring the tank units into the Regular Army establishment. Among the options considered, two stood out from the beginning: first, that the Tank Corps should be a completely separate arm, like the Royal Engineers; and, second, that they should continue under the current system in which officers were temporarily assigned from other branches for specified periods of time.[42] The continued uncertainty over which course would eventually be adopted naturally had produced a severe morale problem among tank officers, and many returned to their original branches rather than risk their careers on so uncertain a future.[43] The matter was finally settled in July 1922, when the Army Council gave formal approval to the establishment of a separate tank corps with its own permanent list of officers and, if the king approved, with the designation "Royal."[44] This permission was given, and in September 1923, the Royal Tank Corps came into being with a peacetime establishment of four tank battalions, one for each of the regular infantry divisions of the postwar army, and eight armored-car companies.[45] This decline in the army's tank strength from a peak of twenty-six battalions to a mere four was disheartening to the tank advocates, but it should not obscure the fact that it actually represented a significant increase in the proportion of tank battalions to the size of the army as a whole when compared to the wartime ratio. In August 1918, Great Britain had eighteen tank battalions in France to support fifty-four divisions, or a ratio of one to three.[46] The Tank Corps as established in 1923 with its ratio of one to one thus represented a threefold increase in the proportion of tanks to the size of the army as a whole. Moreover, this was the same ratio as that proposed by Fuller in November 1918, the difference in actual strength being accounted for by the smaller size of the army as a whole.

The final step in the establishment of the postwar tank forces occurred in 1924, when the army finally adopted a new tank as the mainstay of the Royal Tank Corps. Since the end of the war, it had been experimenting with a number of new vehicles, most notably the Medium D and a new infantry tank, but both vehicles were plagued by mechanical difficulties that could not be overcome.[47] In 1922, the firm of Vickers Ltd. produced a new light tank that was admittedly inferior in design to either of these vehicles, but that was far more reliable. It weighed slightly less than nine tons, had a speed of eighteen miles per hour and a radius of action of 150 miles. Because of its reliability, it was adopted and redesignated the Medium Tank Mark I (but commonly known as the Vickers Medium). It served as the standard medium tank for the British Army until 1937.

The first four years of the life of the Royal Tank Corps, 1923 to 1927, were marked by little progress. In the eyes of the tank advocates, the major reason for this was the appointment of Lord Cavan as CIGS in early 1922. Born in 1865, Cavan entered the Grenadier Guards from Sandhurst at the age of twenty. He served in the Boer War and in World War I and, in March 1918, was given command of all British troops in Italy. His attitude toward change was as conventional as his career. Cavan did not see himself as an opponent of progress, but he firmly believed it must be slow and methodical in order to be of any benefit; radical change was anathema to him. For many, of course, this was a virtue and on the occasion of his retirement the *Army Quarterly* commented:

> He has steadily refused to be hurried into any radical reforms without a thorough and continuous investigation of the proposed changes and neither in the matter of its scientific and mechanical development has he been deflected from his policy by the voice of criticism.[48]

This attitude was naturally frustrating to the tank advocates, who felt that such a deliberate pace was unjustifiable in light of the potential of the new weapon. Fuller, caustic as ever, wrote to Liddell Hart in 1923 that "the mind of this little man is only 800 years out of date! Isn't it wonderful?"[49] What probably annoyed him more than Cavan's caution, however, was the fact that he was having difficulty publishing his books. According to army regulations, a serving officer could not publish any book while on active duty, but could only contribute articles to the military journals. Fortunately, like most senior officers, Fuller had at least a few months between assignments, when he was placed on half-pay, and during these hiatuses he was able to have his books published. This taste of independence proved only more frustrating, however, and after a brief period of half-pay in 1925, he wrote to Liddell Hart, "If they had only let me have six months, I would have blown their heads off. Now I shall have to wait a year or so before I set match to fuse."[50]

The army, however, believed it was progressing at a steady rate and, given the many technical problems of tank design that had to be overcome, at a completely sound pace. At a Royal Academy banquet given in the spring of 1924, Lord Cavan remarked that the army was extensively engaged in research and development in the field of tank design, and he predicted that the end of the year would find England far ahead of any other nation in this field. Reporting the speech, the *Army Quarterly* used these remarks as the basis for criticizing those who called for a more radical program of mechanization. The journal affirmed its belief that tanks would be vital in future war by providing greater mobility to armies and allowing more scope for generalship. It agreed, too, that infantry and cavalry had diminished in importance and that some of their tasks would be taken over by mechanized forces. With equal force, however, it maintained that the older arms would continue to be essential in any future war and that the pace established by the War Office was just right. In short, its position was that the army must by all means go forward, but that it must do so slowly.[51]

Undoubtedly the highlight of Lord Cavan's tenure as far as the army was concerned was the army maneuvers held from September 22 to 25, 1925.[52] Large-scale maneuvers had been a regular feature of the prewar army, as they were the only means by which senior commanders and their staffs received any practical training, but the confusion during demobilization and postwar budget cuts made such exercises a luxury, and this was the first such training since the war. For this event, all four infantry divisions stationed in England plus the cavalry division and two battalions of the Royal Tank Corps were brought together on Salisbury Plain and the surrounding region. After a month of brigade and division training, the units were divided into two opposing forces for a final exercise: the Eastern Force under Lieutenant General Sir Philip Chetwoode (composed of three infantry divisions) and the Western Force under General Sir Alexander Godley (composed of one infantry division and the cavalry division). Each force had one battalion from the Royal Tank Corps.

For the most part, the exercises were strictly conventional, but on the third day General Godley employed tanks as a true maneuver force, the first such use of tanks recorded by any army. In doing this, Godley was applying a minor but highly significant revision of the *Field Service Regulations* of the previous year that stated: "Owing to their mobility tanks may sometimes be employed, especially in conjunction with the cavalry, to make a wide turning movement round the enemy's flank, against his reserves and the flanks or rear of his gun positions and his headquarters."[53] For his attack Godley formed a special mobile force of one tank battalion, a mechanized field-artillery brigade, and a battalion of trucked infantry and ordered them to move around the Western Force's flank and strike its rear. In the approach march the tanks covered thirty-three miles in the dark in just under five hours, but, because their infantry and artillery support fell behind, they had to attack the enemy headquarters alone. Because of this lack of support, the attack was consequently ruled a failure by the umpires, but the operation as a whole lost none of its significance as a result.[54]

Perhaps even more important than this attack was the army's reaction to it. One point of general agreement was that the 1925 maneuvers as a whole had demonstrated the lack of mobility of the older arms, especially when compared to that of the mobile force. Although there was little question that Godley's attack had failed, the problem was seen as a matter of execution, and the army's official report on the maneuvers reasserted the principle established the previous year, declaring that "when our tanks have been improved in type and the numbers have been increased, it is more likely that their employment for action of this type may become an accepted axiom."[55] Until that time, it urged that tanks should continue to be restricted to the direct support of infantry assaults on enemy positions. Thus, while it foresaw greater possibilities for the future, for the time being tanks were still to be used as they had been during World War I.

Little was done during Cavan's last months as CIGS to change this situation. In May 1926, Colonel Lindsay, then inspector of the Royal Tank Corps, prepared a report for the new CIGS, General Sir George Milne, in which he summarized the state of the corps and offered a series of proposals for its future development. The picture Lindsay drew was gloomy. Although the corps had been established three years earlier with a fixed strength of four battalions, it could still only muster two battalions of forty-eight tanks each and this for only a brief period, as spare parts and replacements were lacking. Even worse, its ability to undertake rapid expansion was virtually nonexistent. Because so few tanks were being ordered, the means of production were largely dying out, and Lindsay predicted it would take six to nine months before England would be capable of producing tanks in any quantity. Moreover, there was a severe lack of trained reservists to fill out the corps should expansion be undertaken. Based on the experience of World War I, in which tanks had been shown to be essential for successful offensive action, Lindsay concluded that the army "cannot achieve victory and is ineffective as an instrument of war."[56] To change the situation, he urged that the cavalry be reduced to provide the money needed for mechanization. He admitted that cavalry was still valuable for close reconnaissance and for movement in areas unsuited for armored vehicles, but he felt its functions could be effectively carried out by relatively small numbers of horsemen. Moreover, he argued that combined tank-cavalry operations produced too many tactical and administrative problems to be a viable alternative.[57] In forwarding this report, the QMG commented that "I admit I rather dispair of the views of some mechancial enthusiasts," but Milne's reaction was notably different. He conceded that, in his view, Lindsay's "zeal had outrun his discretion," but he went on to express the opinion that "Colonel Lindsay is on the right lines and we have now to decide how to translate his ideas into action."[58]

The very fact that Lindsay would write such a memorandum to an incoming CIGS, as well as Milne's reaction to it, is indicative of the high hopes the latter's appointment aroused among the tank advocates. Although sixty years old when he assumed the post in February 1926, Milne had expressed views highly progressive for the time, and one of his first actions, the appointment of Fuller as his military assistant, seemed to put the tank people in a key position in the War Office. Liddell Hart, in fact, described Milne in the *Daily Telegraph* as a man "possessed of an exceptional strength of character and decision" and predicted a renaissance for the British Army.[59] Milne proved to be something of an enigma, however, for despite early indications that he would accelarate the process of mechanization, the pace of change did not increase significantly during his tenure. The reasons for this failure are difficult to define precisely, for Milne's sincerity on this matter has not been questioned by any who knew him. It does seem, however, that he lacked decisiveness and, perhaps, even a thorough understanding of the concept of armored warfare that many had initially missed.

Perhaps there is something indicative of the nickname "Uncle George" he had acquired; it reflected a gentle character who lacked the ruthless determination to override lifelong comrades and push through programs that by their nature would cause bitterness and hard feelings. General Broad, who served in the War Office at this time, has recalled that Milne would assign him to a project dealing with an aspect of mechanization, but would become discouraged when faced with conservative opposition and later tell Broad to forget what he had said.[60] An even clearer case of this indecisiveness was reflected in his attitude toward the appointment of Fuller as his military assistant. This step was, of course, an obvious indication that drastic changes were being considered, for, as Liddell Hart wrote at the time, "Colonel Fuller has established himself since the war as the most far-sighted and original military thinker in the country—perhaps, indeed, any country."[61] Yet, Milne seems to have been a bit unnerved by his audacity, and although Fuller's appointment was confirmed in late September 1925, it was not publicly announced for almost two months. In the meantime, as Fuller wrote Liddell Hart, "I know he is a little fearful of his selection. I was not with him a whole day during the maneuvers and I could feel this distinctly." Milne told Fuller not "to say anything about your appointment at present."[62] Milne's tenure as CIGS would be characterized by several bold moves such as this which would not be pursued or sometimes even be restrained.

Disappointment with Milne still lay in the future, however, and the initial optimism that greeted his appointment was given an even greater boost when, less than a month later, the secretary of state for war, Sir Laming Worthington-Evans, announced the formation of a special mechanized force. The tank advocates were naturally delighted by this announcement—shortly after, Liddell Hart compared it to the establishment of the Rifle Brigade at Schorncliffe during the Napoleonic Wars[63]—but in truth the unit was being formed solely to determine the organizational and vehicular requirements of a combined arms mechanized force and not the future of the army as a whole as they hoped. Worthington-Evans went out of his way to spell this out, telling the House of Commons that "the future of mechanicalization is full of difficulties which must be gradually studied before definite decisions are reached." He reemphasized this cautious approach, this time with a note of optimism, when he concluded: "We are, I am advised, already ahead of all other nations in mechanization, and there is therefore no justification for uneconomic haste."[64]

Worthington-Evans's speech gave the impression that the mechanized force would be formed in the summer of 1926, and when this did not occur, it produced charges that the delay was yet another example of the army's lackadaisical or even hostile attitude toward tanks.[65] This, however, was not the case. The army had made no plans to form the unit prior to the secretary's speech, and it was only three weeks afterward that a plan of organization was even proposed for discussion.[66] Furthermore, there were severe equipment shortages that the army could not hope to fill rapidly. In April 1926, Milne informed the secretary of state for war that he hoped to have one tank battalion at its full establishment of forty-eight tanks by the end of the year, but that deficiencies in other types of vehicles were so severe that "the formation cannot be in full working order until 1928."[67] Thus, the fact that the unit was not formed until 1927 was, by all appearances, largely unavoidable and the delay considerably less than originally estimated.

Equipment problems were further complicated by questions concerning organization and administration. In May, the deputy DSD drew up a paper entitled "Scheme for Experimental Work with a Mechanical Force," which went far to reveal the army's state of mind at this time. The report itself was largely technical, ignoring such fundamental questions as the purpose of the force or the assumptions upon which it was being established. Such basic questions apparently did not enter into the author's consideration, who was so concerned with the trees that he did not see the forest. According to his plan, the mechanized force would consist of a tank battalion (the Fifth Battalion, RTC, which he hoped would be at full strength by the next spring), two armored-car companies, a special reconnaissance company, one Field Brigade, Royal Artillery, one Brigade, Royal Artillery (for close support), three infantry battalions, and engineer and signal units.[68] The most significant feature of this plan was that the armored-car companies in the mechanized force would be converted cavalry units rather than RTC armored-car units. He justified this on the grounds that "the main function of cavalry today is protection and reconnaissance and it is only chance which has handed over the armored cars, whose functions are very similar, to the Tank Corps."[69] The tank advocates were chagrined by such thinking. George Lindsay, who, as inspector of the Royal Tank Corps, was asked for his comments, promptly attacked what he considered to be the wasteful diffusion of effort inherent in giving armored cars to the cavalry. Repeating the position he had taken in his private paper to Milne, he strongly urged that all armored fighting vehicles, whatever their specific type, be placed under the control of the Royal Tank Corps. "Experience teaches us," he wrote, "that at the beginning of any new movement it is essential to concentrate all the new effort in one body."[70] He also attacked the incorporation of so much infantry in the force, declaring that one machine-gun battalion of thirty-two guns would be better than the three battalions of sixty-four guns each, as suggested by the deputy DSD. He concluded that "if we are to develop on Machine and Weapon power lines we must cut down on man-power. *We cannot have both.*"[71]

Milne's comments on these differences of opinion are worth recording, for they are another illustration of his own ambivalence on the future of mechanization. He began by declaring, "I consider Colonel Lindsay's ideas very sound," and went on to agree in most details with what he said, but he grounded this agreement in practical terms rather than on principles. He stated his opposition to the idea that "the Tank Corps should swallow the

Army," as some of the tank people were advocating, but he agreed that all armored fighting vehicles should be under Tank Corps control for the present as it had the most experience with them.[72] As a result, the reconnaissance section was placed under the Third Battalion, RTC, equipped temporarily with armored cars. The question of ultimate control of all armored vehicles was left unanswered, however, and would continue to be a major source of conflict between the Royal Tank Corps and the cavalry for another nine years.

During the summer of 1926, little was accomplished in connection with the mechanized force. The units assigned to it gradually began receiving the equipment they needed, and in September the first of them moved to Tidworth, where the force was to be based. The summer was by no means wasted as far as the Royal Tank Corps was concerned, however. There were no large-scale maneuvers as in the previous summer, but the number of tanks available for training was slowly growing, and, overall, the experience gained was probably just as valuable. On July 16, General Godley held a demonstration at Tidworth that pitted an orthodox column of marching infantry against a tank force with cavalry and mechanized support. Liddell Hart, covering the events for the *Daily Telegraph,* gleefully reported to his readers that the latter clearly demonstrated its superiority.[73] More important exercises were held at Aldershot in late August. Here, a battalion of twenty-four tanks made a wide flanking maneuver covering some thirty-three miles to strike at an enemy and, while the attack itself was not judged a success, it was another demonstration that the concept of independent tank maneuvers was gaining wider acceptance. As in the previous year, the maneuver was carried out well, but the actual attack was uncoordinated and poorly executed.[74]

It is clear from these exercises that the army was increasingly recognizing the importance of tanks. Even Liddell Hart admitted at the time that "no one could complain that tanks were neglected in the recent exercises, while even their handling by the various infantry commanders was, with certain exceptions, far more suitable than in any previous year."[75] In its own analysis of the maneuvers,[76] the War Office repeated its criticism of the previous year that independent tank operations were beyond the capabilities of its forces, but it also repeated that this criticism was based on the state of tank development at that time and not on any fixed principle. When Britain's armored forces had improved sufficiently, such tactics would have a valuable place in the army's operations. The memorandum then listed the principles that should govern the employment of tank forces as they then existed: (1) offensively, they should be used primarily to assist the infantry in the capture of its objective; (2) defensively, they should be held in reserve to protect against any deep penetrations; (3) they should never be used before the enemy was located or for tasks of secondary importance; (4) they should always be used in mass as "the dissipation of tanks in small packets is wasteful"; and (5) they should always cooperate very closely with infantry.[77] Clearly, the tenor of this memorandum was to see the tank as a vital weapon of war but one that, in its present state, could only serve as an infantry-support weapon as it had in World War I. Anything beyond this, while admitted to be a possibility for the future, was ruled out for the present.

Thus, by the end of 1926, a full eight years after the war, the British Army had only just begun to think about establishing its tank forces as an independent combat arm. In the eyes of the tank advocates, this was naturally frustrating and, in their view, the result of entrenched opposition to the tank itself. General Broad has recalled that, while he had no proof, he felt certain he was constantly watched when he served as deputy DSD at the War Office in the late 1920s. Although he did not believe his work was hampered, he did believe that those opposed to the expansion of tank arm were determined to be able to anticipate his moves.[78] More important was the army's unwillingness to make significant reductions in other arms to provide more money for tanks. Although the size of the cavalry and Royal Tank Corps were not connected directly except in the minds of many tank advocates, the fact is the army was reluctant to reduce the former except under considerable pressure and this unquestionably reduced the funds available for mechanization. In 1921, for example, the Army Council agreed "in principle" (and only under the insistance of the secretary of state for war) that the cavalry should be reduced by 3,000 men. Yet, three months later the secretary complained that, despite the decision, only 1,200 men had been earmarked for reduction and no actual reduction had taken place.[79]

Yet it is also important to remember that the army training in 1925 and 1926 revealed a willingness on the part of some senior commanders to experiment imaginatively with tanks and the War Office reaction to their efforts was by no means wholly negative. Furthermore, the 1926 report of the Cavalry Committee demonstrated that the army was increasingly becoming aware of the limitations of the horse. Its recommendation that the size of cavalry regiments be reduced slightly to provide funds for experiments with armored fighting vehicles was, while hardly revolutionary, certainly not reactionary.[80] The cavalry had to face other pressures as well. Since the last stages of the war, it had become increasingly difficult to procure both horses and men who knew how to handle them, since the animals were disappearing from civilian life.[81] As a result of these pressures and strict government budgets, the number of line cavalry regiments was reduced from twenty-eight to twenty-one, a reduction of twenty-five percent. Yet this was still insufficient to meet the minimum needs of the Royal Tank Corps.

In the army's view, these measures and the formation of the Experimental Mechanized Force, scheduled to begin tests the following summer, refuted all charges that it was rigidly adhering to the past. Rather, the army viewed its refusal to go as fast as the tank advocates demanded simply as a caution, commonsense approach to a very uncertain future. To do more would run the serious risk of creating an imbalanced army that lacked the combination of stability and firepower as well as mobility necessary to conduct a successful campaign, that is, an army that lacked the ability to conduct a strategy of attrition.

Notes

1. Colonel Sir Hugh Ellis, "Some Notes on Tank Development during the War," *Army Quarterly* 2 (July 1921): 268–69.
2. Colonel J. F. C. Fuller, "The Application of Recent Developments in Mechanical and Other Scientific Knowledge to Preparation and Training for Future War on Land," *Journal of the Royal United Service Institution* 65 (May 1920): 251–61.
3. Anthony Trythall, *"Boney" Fuller* (New Brunswick: Rutgers University Press, 1977), 87.
4. Fuller, "The Application of Recent Developments," pp. 249–50.
5. Colonel J. F. C. Fuller, "The Influence of Tanks on Cavalry Tactics," *Cavalry Journal* 10 (April 1920): 109–32; (July 1920), 307–322; (October 1920), 510–530.
6. Ibid. (April 1920): 111.
7. Ibid. (October 1920): 528.
8. Ibid. (April 1920): 109; (July 1920): 307.
9. W. D. Croft, "The Application of Recent Developments in Mechanics and Other Scientific Knowledge to Preparation and Training for Future War on Land," *Journal of the Royal United Service Institution* 65 (August 1920): 467.
10. S. R. Mason, "The Reorganization of the Royal Regiment of Artillery with a View of Obtaining in War the Best Value from Modern Equipments with a Minimum Expenditure of Human Energy," *Journal of the Royal Artillery* 47 (July 1920): 139–55; with the same titles, the article by W. H. Moore appeared in the August 1920 issue, pp. 199–212 and that by A. L. Pemberton in the September 1920 issue, pp. 215–64.
11. The fourth essay, by R. Hilton was published in the October 1920 issue, pp. 317–32; the fifth, by R. G. Cherry, in the November 1920 issue, pp. 361–78.
12. P. Johnson, "The Use of Tanks in Underdeveloped Countries," *Journal of the Royal United Service Institution* 66 (May 1921); 191–204.
13. Ibid., p. 203.
14. C. B. Dashwood-Strettell, "Cavalry in Open Warfare Illustrated by the Operations Leading up to the Occupation of Mosul in November 1918," *Journal of the Royal United Service Institution* 66 (November 1921): 598–617.
15. Anonymous, "Auream Quisque Mediocritatem Diliget," *Army Quarterly* 5 (October 1922): 122–29.
16. Liddell Hart advocated a similar system of progression in "The Development of the 'New Model' Army: Suggestions on a Progressive but Gradual Mechanization," *Army Quarterly* 9 (October 1924): 37–50.
17. See above, pp. 22–27.
18. Brigadier General G. A. Weir, "Some Critics of Cavalry and the Palestine Campaign," *Cavalry Journal* 10 (October 1920): 534.
19. Great Britain, Parliment, *Hansard's Parliamentary Debates* (Commons), 3rd series, 89 (18 March 1921): 1354–55.
20. Anonymous, "Editorial," *Army Quarterly* 8 (July 1924): 255.
21. Major B. C. Denning, "Military Prize Essay, 1929," *Army Quarterly* 18 (July 1929): 235–57.
22. Great Britain, War Office, *Field Service Regulations,* vol. 2: *Operations* (London: HMSO, 1920), pp. 31–32. The 1924 edition repeated these points in virtually the same words.
23. Ibid., pp. 185–86, 226. The same points were made in the 1924 edition, pp. 121–22, 144, 194.

24. Ibid., pp. 22–23. The same points were made in the 1924 edition, pp. 14–15.
25. Ibid., pp. 179–80. The same points were made in the 1924 edition, p. 123.
26. Ibid., p. 141. It was repeated in the 1924 edition, p. 36, except for the passages in brackets, which were deleted, and the passage in parentheses, which was added in the later edition.
27. See above, pp. 65–67.
28. Victor Germains, *The Mechanization of War* (London: Praed and Co., Ltd., 1927), pp. 162, 218–19.
29. Lieutenant General Sir Charles Broad to Rickard, June 23, 1970, Broad Papers, Center for Military Archieves, Kings College, University of London.
30. Germains, *The Mechanization of War,* pp. 57–77, 164–66, 222.
31. Ibid., pp. 160–61, 177, 179, 228–30.
32. Ibid., pp. 152, 175–76, 181–82, 212–13.
33. Ibid., p. 178.
34. Ibid., pp. 234–43.
35. Anonymous, "Editorial," *Army Quarterly* 16 (April 1928): 178–79.
36. Anonymous, "Army Reorganization," *Journal of the Royal United Service Institution* 63 (May 1918): 286–300; "Clericus," "The Army after the War," *Journal of the Royal United Service Institution* 63 (August 1918): 493–99.
37. Colonel J. F. C. Fuller, *Tanks in the Great War* (London: John Murray, 1920), pp. 163–64; Major General Sir C. H. Harrington, "The Passing of an Army from War to Peace," *Journal of the Royal Artillery* 47 (May 1920): 49. In addition, the British also equipped and trained one Canadian and three American tank battalions.
38. Memoranda of 15 November 1918, and December 1918, Box 1, Fuller Papers, Special Collections Department, Rutgers University Library, New Brunswick, New Jersey.
39. "The State of the Tank Corps," an undated memorandum attached to a letter from the DSD dated 31 March 1919, Fuller Papers.
40. Memorandum of 5 April 1919, from the deputy CIGS to the secretary for war, Fuller Papers.
41. WO 163/25, "Minutes of the Proceedings of, and Precis Prepared for the Army Council, 1920–1937" meetings of 1920, pp. 125–26.
42. B. H. Liddell Hart, *The Tanks,* 2 vols. (New York: Frederick A. Praeger, 1959), I:202.
43. George Lindsay, "Report on the Royal Tank Corps," an unpublished paper dated December 1927. Lindsay Papers, States House, Medmenham Marlow, Buckinghamshire.
44. WO 163/28, Meetings of 1922, pp. 26–27.
45. Liddell Hart, *The Tanks,* p. 225. Richard N. Ogorkiewicz, *Armoured Forces* (London: Arms and Armour Press, 1970), p. 149. The armored-car companies of the Royal Tank Corps served in India throughout the interwar period.
46. Brigadier General Sir James E. Edmonds, *Military Operations France and Belgium: 1918, August 8–September 26* in *History of the Great War Based on Official Documents,* Brigadier General Sir James E. Edmonds, gen. ed. (London: MacMillan and Co., Ltd., 1947), contains the complete Order of Battle for the BEF in France on 8 August 1918 in Appendix 1, pp. 521–34.
47. In his March 1919 memorandum on the state of the Tank Corps, Fuller reported that the Medium D was in the first stage of development and that it would be ready for production by May 1. Box 1, Fuller Papers. Yet in the October 1920 meeting of the Army Council, the MGO reported that the tank's continued mechanical problems had led to considerable cost overruns but that he expected it to be ready for production by April. WO 163/25, pp. 125–26.
48. Anonymous, "Editorial," *Army Quarterly* 12 (April 1926): 8.
49. J. F. C. Fuller to B. H. Liddell Hart, 12 November 1923, Liddell Hart Papers, States House, Medmenham, Marlow, Buckinghamshire.
50. Fuller to Liddell Hart, 28 September 1925, Liddell Hart Papers.
51. Anonymous, "Editorial," *Army Quarterly* 8 (July 1924): 226–29; and 10 (April 1925): 1–3.
52. This account is based on Liddell Hart's articles in the *Daily Telegraph,* 31 August to 28 September 1925.
53. *Field Service Regulations* (1924), p. 123.
54. *Daily Telegraph,* 25 September 1925.
55. *Daily Telegraph,* 26 April 1926. The report was not issued until this time but was pre-

pared under Lord Cavan. For an informal, unofficial view see Anonymous, "Editorial," *Army Quarterly* 11 (January 1926): 231–32.

56. Major General George Lindsay, "Suggestions Regarding the Best Means of Rendering the Royal Tank Corps in Particular and the Army in General More Suited to the Probable Requirements of Future Warfare," an unpublished paper dated May 1926; Lindsay Papers, 2–3, 36–37.

57. Ibid., pp. 9–10.

58. Ibid. The QMG's comments are dated 28 June 1926. Milne's comments are undated.

59. *Daily Telegraph,* 19 February 1926.

60. Interview with Lieutenant General Sir Charles Broad, Beaminster, Dorset, England, 19–20 August 1970.

61. *Daily Telegraph,* 23 November 1925.

62. Fuller to Liddell Hart, 28 September and 3 October 1925, Liddell Hart Papers. Fuller was officially appointed on 25 September 1925, effective 19 February 1926. The letters are in Box 1, Fuller Papers. The appointment was announced in the *Daily Telegraph* on 23 November 1925.

63. *Daily Telegraph,* 1 September 1926.

64. Great Britain, Parliament, *Hansard's Parliamentary Debates* (Commons), 3rd Series, 193 (15 March 1926), p. 78.

65. In the spring of 1927, Liddell Hart reported that nothing had been done in the field of mechanization during the previous year. *Daily Telegraph,* 22 April 1927. Years later, he wrote of how the summer of 1926 "slipped away" while the army bureaucracy worked out the details of the force. *The Tanks,* 1:243.

66. WO 32/2821, "Proposed Establishment of the Experimental Mechanical Force," minute 2A, dated 8 April 1926.

67. Ibid.

68. W. O. 32/2820, "Proposals for Initiating the Experimental Mechanical Force," Minute 1A, p. 1.

69. Ibid., p. 3.

70. Ibid.

71. Ibid.

72. ibid.

73. *Daily Telegraph,* 19 July 1926.

74. *Daily Telegraph,* 28 August 1926.

75. *Daily Telegraph,* 24 September 1926.

76. These were contained in the "Memorandum on Army Training: The Collective Training Period, 1926," Liddell Hart Papers.

77. Ibid., p. 10.

78. Interview with General Broad.

79. WO 163/26, "Minutes and Proceedings of and Precis Prepared for the Army Council, 1921," meeting of 21 February 1921, p. 10; and meeting of 25 May 1921, p. 118.

80. See above, pp. 25–27.

81. This problem was first mentioned in E. A. P. Hobday, "Horse Management in the Field," *Journal of the Royal Artillery* 45 (May 1918): 41–59.

• 5 •

The Years of Experiment: 1927–34

The period between 1927 and 1934 was the turning point for the British Army and the theory of armored warfare during the interwar years; unfortunately, it was a turning point, to paraphrase A. J. P. Taylor, at which the British Army failed to turn. There were many significant advances for the Royal Tank Corps itself and for the new strategic and tactical principles that were being proposed for its employment. The corps was expanded from a single battalion to a full brigade of four battalions. Its training was greatly enhanced after 1929 by the publication of the first manual for tank operations issued by any army; henceforth, tank companies and battalions would be able to maneuver just as effectively as their infantry and cavalry counterparts. And, finally, four of these eight years witnessed extensive training exercises and trials designed specifically to test the capabilities and principles of an armored force operating independently. So significant were the cumulative effects of those advances that by 1934, the British Army had the largest, most tactically advanced, and best-trained tank arm in the world.

In the end, however, the disappointments and failures of this period weighed more heavily than these advances. After 1928, the lively public debate over the future of mechanization all but disappeared from the military journals. Moreover, the retirement of J. F. C. Fuller in 1933 removed the most advanced theorist of armored warfare from the ranks of the army. Although he retired with the rank of major general and could not claim to have been unrewarded for his services, he never returned to the Tank Corps after the war, and he played no role in the training exercises of this period. Most important, however, was the fact that none of the advances during this period swayed the army from the conviction that the strategy of attrition was completely sound or that tank forces could alter this strategy in any significant way. By the end of 1934, there is no question that the army had accepted tanks as a powerful weapon not only for infantry support, but also for highly mobile independent operations; but the theory of armored warfare had been firmly rejected, and the pattern of tank development for the remainder of the interwar years was established.

The Experimental Mechanized Force

During the winter of 1926–27, the tank advocates' hopes ran high as preparations for the Experimental Mechanized Force (or EMF) proceeded steadily. Gradually, new equipment was distributed to the units comprising the force and they completed their move to Tidworth. The tank advocates' hopes were heightened even further when Fuller was appointed to command the force on December 31. Although Fuller was apprehensive at the appointment—he wrote Liddell Hart, "I am by no means overjoyed as it is a first day of creation show and I am not in a position to emulate the Almighty"[1]—he accepted the position nonetheless. At first appearance it was an ideal situation: a revolutionary new unit commanded by the army's most advanced theorist of tank warfare. Unfortunately, these hopes were soon dashed and the prospects of the EMF crippled even before the force came into existence.

The problem that proved to be insuperable was the command arrangements under which Fuller was to operate. The previous June, the War Office had decided that the command of the force would be placed in the hands of the colonel commandant of the Seventh Infantry Brigade with the necessary supporting units attached to it. However, these supporting units would each be responsible for its own internal training and come under the brigade commander only during the final exercises.[2] This situation was obnoxiously complicated, to say the least, for it not only made coordinating training for the entire force difficult, but it also meant that the commander would have the additional responsibility for the garrison of the entire Seventh Infantry Brigade, forcing him to divide his own duties. The situation was exacerbated by the fact that the force would not be directly under the War Office as Fuller desired, but, rather, under two intermediate headquarters the Third Division and Southern Command. Fuller's opinion of the officers holding these respective commands, General Sir John Burnett-Stuart and General Sir Archibald Montgomery-Massingberd, was not particularly high and he no doubt feared their interference. Fuller learned of these arrangements by January 1927,[3] and although he accepted the appointment, he immediately set out to correct the situation. On February 8, he wrote a long letter to Milne stating his objections to the terms of his appointment. He first noted that the Seventh Infantry Brigade consisted of four battalions, only one of which would be engaged in the experiments. This meant, in effect, that he would be the commander of two units, one experimental and the other conventional, and would be unable to concentrate his attention on either one. Moreover, he argued, the staff assigned to assist him with the mechanized force itself was totally inadequate. Noting that a normal infantry brigade had only four units, all with identical organizations, he reminded Milne that the mechanized force would contain ten units, only three with established organizations and equipment. As a result, he would not only have to test the unit as a whole, but would also have to test each of its components. Yet, he wrote, "I am given only the normal Brigade Staff to assist me." Finally, Fuller complained about having to assume command of the Tidworth garrison as well as his other duties and once again that he had been given inadequate staff. He concluded:

> I feel strongly that my duty is to make this formation a success, and, as will be realized, if it is not successful every soldier opposed to modernization will at once proclaim the newer arms valueless.

To correct this situation, Fuller requested that (1) he be given a second in command to take charge of the three nonexperimental battalions of the Seventh Brigade; (2) the cavalry brigade commander or an administrative commandant be appointed to command the Tidworth garrison; (3) he be given an additional permanent staff captain to assist in the running of the mechanized force; (4) at least one of his clerks be a good stenographer; and (5) he be given a small entertainment allowance since he had to deal with such a large number of units.[4]

Fuller's demands are peculiar in a number of respects. There is no doubt that the arrangements set forth by the War Office were less than ideal, but it is difficult to imagine that it would not have been within Fuller's power to mitigate if not correct the worst of these deficiencies once he assumed his command. As a full colonel with the authority of both a brigade and a garrison commander, it would have been a relatively simple matter for him to delegate some of his responsibilities and to appoint additional personnel to his staff for as long as they were needed. Certainly, for an officer of his rank and position to bother the CIGS with such petty details as obtaining a trained stenographer for his staff was more than a little ridiculous. But Fuller did not see it that way. His pride and intellectual arrogance had long since alienated him from many of his fellow officers, and, as he told Milne, his failure in this venture would be greeted with considerable joy among these men. Hence, the command arrangements for the experimental force had to be entirely sound and clearly established in advance lest he find himself hamstrung or subject to charges of somehow acting improperly at a later date. Fuller, in short, had been outside the "club" for too long to expect it to come to his aid now; his orders had to be precise because he had no other source of protection.[5]

The ultimate tragedy of this affair is that Fuller's assessment was probably wrong. The War Office records regarding the command arrangements contain no evidence that they were designed to inhibit the potential of the experimental force, although it is highly unlikely that such considerations would have been put into writing. It is improbable, however, that Milne, for all his caution, would have tolerated such an arrangement. More likely, the War Office simply considered these command arrangements to be the most logical available. Since the mechanized force was only a temporary unit, the commander should be given a permanent position and the Seventh Infantry Brigade was the logical choice since a portion of it was to be used in the tests. As for placing the EMF under the Third Division and Southern Command, again this seems to have been based on standard army procedures, as the exercises were to be held in the jurisdiction of these commands. This procedure also was followed in 1935 and 1936 for additional mechanization experiments and again seems to have been based on the logic that, since the tests were being held within the bounds of a major command in the British Isles, the commanders of those areas should exercise an overall supervisory function.[6] That Fuller's position may have been difficult—especially in his relations with Montgomery-Massingberd, who disliked him—is undeniable, but it was not created to inhibit him nor would it had he used a little discretion. No doubt Charles Broad, who was deputy DSD in the War Office at this time, was right in his conclusion that "he [Fuller] should have gone on and taken it."[7]

In any event, Fuller's first appeal to Milne of February 8 was rebuffed with a curt "Don't be silly."[8] Following another appeal a week later, Fuller wrote Burnett-Stuart, hoping that he would intercede with the CIGS. Rather than provide support, however, Burnett-Stuart chided Fuller for his unwillingness to accept an admittedly less than ideal situation and failure to realize the distinction the appointment carried. "There are many first rate fellows," he warned, "who would give their eyes to have it, so don't think you are conferring a boon on society by accepting it." After describing some of the problems they would face, he told Fuller that no one was expecting miracles and that he, for one, was pleased to see Fuller get the appointment:

> I confess I am looking forward very much to having you work with me; and incidentally I applaud a system which insures that individuals containing a spark of the divine light should take their places among the fighting troops and help to vitalize the organization from within. It may be a less spectacular role, but it is infinitely more useful than pummeling from above.[9]

This was hardly the kind of advice Fuller expected or wanted.[10] For two weeks he held his own council, but on March 3 he was galvanized into action when the secretary of state for war announced to the House of Commons that "arrangements have been made to assemble in the Salisbury Plain area an experimental mechanical force, composed entirely of mechanized units . . . whose primary function will be to study the tactical employment and the organization of a highly mobile force of this nature."[11] This was a flat contradiction to the terms of his appointment—there was no mechanized force as such, but only a tank battalion with other units to be attached for a short period—and Fuller again went to Milne to have the matter clarified. The CIGS again refused to consider any changes, and Fuller returned to his office and wrote his resignation from the army.

Fuller's act rocked the War Office—in his own words, it "fell like an exploding bomb."[12] Burnett-Stuart wrote him expressing the hope that "my letter did not contribute to it in any way."[13] More important to Fuller was an appeal from Lieutenant General Sir David Campbell, until recently the military secretary to the CIS. and now commander at Aldershot, who urged him to withdraw his resignation. On March 29, Fuller acceeded to this request, exlaining to the new military secretary that he had submitted the resignation only because he wanted to avoid the appearance of bargaining over the post. He then undermined this explanation by expressing his willingness to accept the command if it was offered again, providing that it met the description made by the secretary of state for war. This second letter was accepted, but there was no mention of his returning to the mechanized force. On April 9, his appointment to command the Seventh Brigade was canceled and on April 27, he was offered the position of GSO1 (General Staff Officer, 1st grade) to the Second Division at Aldershot.[14]

Liddell Hart, who had remained quiet thus far, broke the story to the public on April 22, the day following the appointment of Colonel R. J. Collins to command the Seventh Brigade. In an article entitled "An Army Mystery: Is There a Mechanized Force?," he recalled Worthington-Evans's speech of March 3 to the House of Commons and another statement four days later that the force would be commanded by an expert in mechanization. "Yet," Liddell Hart wrote, "only six weeks have passed and an announcement next comes which is in flat contradiction of all their points." He concluded:

> Has the scheme broken down, or was the formation of such a force no more than a figure of speech? Parliament and the public, already uneasy over the ill-conceived affair of the Territorial bounty, have a right to enlightenment, which will be awaited with some anxiety.[15]

Liddell Hart's disclosure threw the War Office into its second turmoil in as many weeks. Worthington-Evans, who had never been informed of Fuller's resignation, was particularly incensed and accused his advisors of misleading him. Milne, in self-defense, tried to shift the blame to Burnett-Stuart for failing to come up with a workable arrangement.[16] Within a week, however, a new arrangement was devised. The War Office announced that a distinct mechanized force was being formed at Tidworth under the command of Collins, who would also command the Seventh Brigade. This arrangement—unlike Fuller's, which simply assigned him to the Seventh Brigade with other units to be attached for the purpose of the tests—was more to Liddell Hart's liking. In reporting the announcement, he offered his congratulations to the army, but seemed to imply that he would be watching to make sure that nothing was done to weaken the intent of these decisions. The problem of Collins's dual command still bothered him, however, and on May 12, he tried to repeat his earlier success in applying public pressure on the War Office. In an article appearing on May 12, he noted that the appointment of an officer to simultaneous commands violated both common sense and established military principles and caustically concluded: "In view of its disregard in present circumstances, it seems that we must coin a new phrase—that principles are made for convenience, and not for consistent observance."[17] This time Liddell Hart's tactics failed; the War Office refused to budge and Collins remained in both positions throughout the existence of the mechanized force.

The repercussions of this affair were grave both for Fuller and Liddell Hart personally as well as for the chances of the British Army's adopting the theory of armored warfare. For Fuller, it marked the end of his career as a respected voice in the army. Milne apparently felt he had been betrayed by Fuller, and although there were cordial contacts between the two afterwards, they were limited and the CIGS never again indicated an interest in employing Fuller's experience in mechanization.[18] Fuller served in a variety of posts until late 1930, when he was promoted to major general and placed on half-pay. He never served on active duty again. As for Liddell Hart, he temporarily lost his privileged position in the War Office whereby he had easy access to inside information and, apparently, the ear if not the sympathy of some senior officers. His remark in the May 12 article regarding the violation of principles was especially antagonizing to many and prompted Montgomery-Massingberd to write him,

> You give the impression of a young man in a hurry who wants to make a literary name rapidly by slashing articles and a certain amount of advertisement and expects criticism of senior and experienced officers who are having an extremely difficult time, to be taken lying down.[19]

In the final analysis, however, it was the EMF that suffered most from this affair. Fuller, with all his idiosyncrasies, was still the only officer in the British Army at this time who had the knowledge, experience, and rank to train and lead the EMF in a way that could have demonstrated the potential of armor. Whether or not he could have convinced the army at large of the correctness of his views is not certain—as we shall see, the results of the 1931 and 1934 exercises are not encouraging in this respect—but without him the force was doomed to a crippled existence for the start.

The Experimental Mechanized Force formally came into existance on May 1, 1927. It consisted of the Fifth Battalion, RTC, now fully equipped with forty-nine Vicker's Medium tanks, the Third Battalion, RTC, with two armored-car companies and one company of newly developed "tankettes," a field brigade of heavy artillery, a battalion of light artillery, a company of engineers, and a machine-gun battalion.[20] Its unique nature lay not only in the fact that it was wholly mechanized, but also in that it was composed of all branches of the army; it was, in reality although not in theory, an opportunity to see the army of the future in miniature form, to discover some of the problems it would face, and to get some ideas of the lines along which weapons and organization should develop. Unfortunately, its ability to accomplish these tasks was marred from the beginning by the army's inability to provide all the neccessary vehicles in adequate numbers. This deficiency was not readily apparent because the tank battalion was up to full strength, but many of the other units were forced to use vehicles unsuitable to the units' missions and much dislocation naturally resulted.[21] In an effort to solve this problem, Collins divided the force into three groups—fast, medium, and slow—and while this helped the situation, it by no means solved it and led to conclusions that were to have dire results for the development of Great Britain's armored forces in the 1930s.

The actual training of the EMF as a whole began on August 19, following a summer of individual unit training. It opened with a series of brief exercises designed to ascertain the basic capabilities of the force.[22] The first two required it to move under noncombat conditions in order to test march discipline and communications, and, in the second, conducted at night, cross-country movement without lights. Although both exercises were far more complicated on the ground than they appeared on paper, they were made more difficult by the conditions under which they were conducted. The day march, for example, was not really under peacetime conditions, as the force had to contend with roadblocks and air interdiction. During the night exercises, the weather conditions were so severe that even Liddell Hart expressed surprise that the entire tank battalion was able to complete the march successfully. The third exercise was designed to test the ability of the force to protect itself while both on the move and encamped. The last two exercises were designed to test the offensive capabilities of the force, both in locating and destroying a conventional force of all arms, and then against an enemy raiding party.

All of these exercises were naturally covered by Liddell Hart in great detail. His views on mechanization were widely known by this time, and he used his position unabashedly to aid the cause in which he believed so strongly. *Daily Telegraph* readers were treated to extensive accounts of the operations under such headlines as "Tanks in Action on Salisbury Plain," "Bombing Attack Thrills," and "Triumph for Mechanized Army."[23] Beneath this sensationalism, however, Liddell Hart's ability as an acute observer and analyst did not fail him. There was little doubt in his mind that the tests were demonstrating not only the feasibility of a mechanized force, but also the advantages it would confer on an army on future battlefields.

Yet there was also distinct disappointment. The mechanized force, he wrote, continued to be hampered by those who did not see its full capabilities, so that, while the exercise demonstrated some of its potential, the full measure of its strength was yet to be realized. Indeed, he went so far as to declare that "it is questionable whether it [the EMF] has attempted anything that would not be done by ordinary troops of a division of the present pattern."[24] True, it could do these things more efficiently, but the goal of the tank advocates was not to improve on the existing methods of the conduct of war—this was clearly impossible—but to change the nature of war radically, to throw out the old pattern completely, and to start anew. In this sense, they viewed EMF as a failure.

Undoubtedly, the highlight of the training season as far as the EMF was concerned was General Milne's visit to the camp on September 8. After viewing the training, he spoke to the officers of the force, giving a speech that Liddell Hart called "a hallmark on the progressive thought and a probable landmark on the evolution of our army."[25] Although the speech itself contained nothing that Fuller or Liddell Hart had not already written—parts of it, in fact, seem almost to paraphrase Liddell Hart's *Paris; or the Future of War*—the fact that these ideas were being expressed by the CIGS gave them a significance that was lost on no one. Milne opened by praising the work of the EMF, calling it "extraordinarily good" and frankly admitting, "I never thought for one moment that you would get so far." He then outlined the basic problem of war at that time: the stagnation produced by the tremendous firepower of the machine gun. It was the purpose of the EMF, he explained, to find a solution to this problem, and he stated his belief that they had not only accomplished this, but in so doing, also had set the course for the future. They had shown that tanks would be the most effective antidote to air power and gas, and he even predicted the development of armored divisions for massive long-range strikes against enemy forces. This was a significant feat and he admitted that "I was a little nervous of the experiment but the zeal that everybody had shown has helped us enormously because I am perfectly certain that we are working on absolutely the right lines." True to his ambivalent nature, however, Milne could not leave without expressing a degree of caution. He warned that complete mechanization was still a long way off—he said he would only see the very beginning of it while CIGS—and he cautioned against those who felt the Royal Tank Corps would absorb the rest of the army. There was, he said, a permanent need for infantry and cavalry, not only for defensive purposes in a major war, but also for use in the empire. In short, tanks would play an increasingly important role in warfare and he would do all he could to lead the way, but the process would be long and slow, and this must be accepted by the tank advocates just as the growing role of the tank would have to be accepted by those in the infantry and cavalry.

The aftermath of Milne's speech was another example of his habit of taking bold steps and then retreating. After his speech, he expressed his intention of having it issued publicly to the army and gave Liddell Hart permission to publish a summary in the *Daily Telegraph.* When he returned to the War Office, however, he was apparently beset by conservatives, who persuaded him that this would not be wise. Fortunately, Liddell Hart's summary was published before Milne's second thoughts occurred, but beyond that, the speech was a dead letter. As Fuller told Liddell Hart at the end of the year, "The CIGS's address has not been published. As far as we are concerned it has been locked up and nobody knows anything about it."[26]

Following Milne's speech, a final exercise was held on Salisbury Plain on September 15 and 16. This was the largest exercise of the year and involved not only the EMF, but also the Third Division, a cavalry brigade, and units from the RAF. The forces were initially placed approximately eighty miles apart, and the mechanized force was given the task of penetrating the enemy's cavalry screen and attacking his infantry while avoiding aerial discovery. This was easily done. The force moved far more rapidly than expected and was able to penetrate the cavalry screen and box in the infantry so that it could be subject to devastating air attacks. The exercise was over, however, before the final act could be completed.[27]

The ultimate significance of these tests, however, lay less in what they actually accomplished than in how interested parties in and close to the army interpreted the results. In this there was so great a variety of opinions that it is impossible not to wonder whether everyone was commenting on the same exercises. Speaking for the tank advocates, Liddell Hart was naturally optimistic about the course of the exercises. In his *Daily Telegraph* articles, he called 1927 a "Year of Demonstration," declaring that it had shown beyond all doubt that the British Army as it then existed was not merely obsolescent but obsolete. The mechanized force, he declared, could strike when and where it wanted and was virtually invulnerable to attack. Even aerial observation was only of limited value unless the force was checked at some point. The problems that remained were primarily ones of technique rather than of concept; they could be worked out by either training or, perhaps, a change in design or new equipment. The most important problem still to be overcome was that of tactical control. Because the force moved with such great speed (for the time) and over such large distances, this was an especially acute problem. It was one that could be solved, however, with a little time and work. The only complaint he registered with regard to the conduct of the exercises was that the umpires vastly overrated the effectiveness of the antitank gun. The superior mobility of the tank, he felt, would definitely give it the edge.[28]

Liddell Hart was by no means the only one speaking out on the subject, and it is a mark of the interest the exercises generated that, during the fall and winter of 1927, numerous articles appeared in all of the military journals written both by men directly connected with the exercises and by independent observers. In December, Collins gave a speech before the Royal Artillery Institution notable more for what it revealed about his limited view of armor than anything else. It contained nothing about the overall concept of armored warfare or of the effectiveness of mechanized forces against conventionally equipped forces. Indeed, it is questionable whether he even believed mechanization would have any real impact on the conduct of military operations. Rather, he emphasized that the exercises had not so much answered any questions about the future of the army as that they had helped to define some of the questions it faced more clearly. He was particularly concerned with the vulnerability of the force against enemy artillery and antitank weapons and clearly felt that these would be able to restrict the effectiveness of the force to a considerable extent. The possibility of using the EMF's speed as its chief protection as the tank theorists advocated passed him by completely.[29]

A more interesting speech was given by General Sir John Burnett-Stuart, who, as commander of the Third Division, had exercised a general supervisory role over the training.[30] Hardly one of the tank advocates, Burnett-Stuart nonetheless saw more clearly than Collins the possibilities of tank warfare in the future. He was well aware of many of the financial and technical problems facing the army and used them to justify a "go slow" policy. Yet, the exercises had convinced him of the potential power of such a force. He was now convinced that tanks should no longer be regarded merely as an infantry-support weapon and that the time had come for them to move out "as a principle, not merely as an assistant." He was well aware that this meant a reduction in the importance of the infantry and cavalry, but he believed this was inevitable. He agreed that there were certain "tank-proof" areas—locales in which tanks simply could not operate because of the nature of the terrain—but maintained that such areas were never decisive in war and should therefore not be a major consideration. Nonetheless, he conceded that the infantry and cavalry would be around for a long time because financial considerations, the enormity of reorganizing the army as a whole, and the need for imperial cooperation prevented rapid and drastic change.

Most of the other articles in the journals reflected one or the other of these views: either they agreed with the basic concept but saw a long process of change for financial or technical reasons, or they were so concerned with the technical problems that they failed to perceive the possibilities of the basic concept.[31] It was quite different with the army's official evaluation of the exercises contained in an army training memorandum issued the following spring. Indeed, the memorandum gives the impression of having been written specifically to dampen enthusiasm for the force. It opened with a brief statement upholding the idea that freedom of thought within the army was necessary to achieve progress, but, having said that, it then imposed guidelines that virtually made freedom of thought impossible. In particular, the memorandum declared that, while such freedom was good, there was a need to "impress upon all officers the necessity for cultivating a sense of proportion." It then pointed to certain principles of tactical training that had been violated during the summer's training—such as the failure to secure proper fire support before assaulting a position—and called on all officers to ask themselves before taking an action, "How far does the action I am taking conform with correct principles?"[32] Superficially these remarks made considerable sense. The need for proper fire support before infantry assaults was certainly a point of critical importance, and the failure of several units to ensure that this was being done had been noted by Liddell Hart in his own comments on army training that year.[33] The tenor of the memorandum, however, seems not to have been directed at correcting failures in tactical training; rather, it seems to have been designed to instill a reverence for the principles of military action as established by the hierarchy. In his comments on the use of inadequate firepower, for example, Liddell Hart had argued that it simply made bad sense and that it would not work in actual combat. The training memorandum, however, based its criticism primarily on the fact that it violated principles.

This in itself might not have been important except that the same spirit was carried over in the memorandum's assessment of the EMF. It firmly put down any hope that the EMF was the germ of the future army. Rather it declared,

> In this experiment no revolutionary change in the organization of the British Army is being aimed at. The object in 1927 is to ascertain the best organization and method of employment of such mechanized forces as we may hope to have at our disposal in the next few years and to try out the various forms of vehicle available.[34]

Accordingly, the memorandum stated that such forces might be employed for strategic reconnaissance and wide flanking movements while the enemy was pinned down by the friendly main force, thus recognizing the value of the independent use of armor. But it did not go beyond that point. It only concluded that the exercises had provided "valuable data"—which were not specified or even hinted at—for future guidance and that the other branches of the army should not overreact against the possible threat of such a force:

> We must not allow the threat of action by a mechanized enemy to upset all our preconceived notions of war. Given an infantry anti-tank weapon and provided we have a proper conception of how to handle our artillery, and provided, further, that we study how to use ground and how to use our reconnaissance units, aeroplanes, armoured cars, etc., there is no reason why the presence of a mechanized force should strike terror into formations and commanders.[35]

Thus, while the tank advocates saw the EMF as an instrument of revolution, the War Office saw only a potentially valuable addition to the traditional arms operating according to well-established principles.

In the spring of 1928, the armored force began its training once again. It had a new name—in his speech of the previous September, Milne had rechristened it the Experimental Armoured Force—but its organization and equipment were the same. As in the previous year, the force began with the training of its subunits, but it quickly became involved in another series of demonstrations. On March 21, a demonstration was held for the public to honor the visiting rulers of Afghanistan. In July, two more were held, the first for the students at the Staff College and visitors from the Indian Army and the second for members of Parliament. These demonstrations, needless to say, were designed to impress the spectators with the power of the army rather than to provide training experience, and, while they involved nothing new in the development of the armored forces, they were valuable propaganda.[36]

The training of the Armoured Force as a whole resumed on August 20 and consisted of a series of six exercises, each lasting for only one or two days, and a large Southern Command exercise held from September 19 to 22.[37] The purpose of each of the brief exercises was to test the ability of armored forces to operate under various tactical situations and the ability of conventional units to deal with them. Undoubtedly the most significant of the exercises was the third, during which the force was ordered to strike at the enemy's lines of communication while the main body of the army struck at his main forces. The Armoured Force moved better in this exercise than in any other and was able to ford a river in the face of enemy opposition and neutralize much of his rear area. The effectiveness of this attack, wrote Liddell Hart in the *Daily Telegraph,* demonstrated that this was the true role of an armored force, but he voiced strong objection to the use of the word *raid* to describe it. He felt such a term vastly underrated the effectiveness of such operations, placing them on the level of the cavalry raids of the American Civil War, which had little direct affect on the outcome of battles, and he asserted the use of armored forces in this manner came closer to what Turenne or Napoleon had called the "rear attack." Such operations required very close cooperation between the mobile force and the main body of the army, and they required that the former be not merely a small portion of the army, but a major component that carried a decisive impact of its own. His own positon was that "the armored force must be expanded to effective proportions. If it is kept as a mere toy, training in proportion to the whole, it will not have the power to achieve more than raids, damaging, but hardly decisive." This, he concluded, was the difference between the concept of cavalry operations in the nineteenth century (which the army was coming to apply to tank forces) and the strategy of tank warfare (as advocated by Fuller and himself).[38]

This distinction is vitally important for understanding the further development of both the strategic doctrines and the armored forces of the British Army. Both concepts advocated the independent use of armored formations to maximize their range and mobility, and the fact that the army was moving in this direction was a significant change that cannot be ignored. But the two concepts were still far apart. The cavalry concept of tank operations was, quite simply, based on the strategy of attrition updated by modern equipment. The use of the word *raid* to describe armored strikes into the enemy's rear indicated the army's continued belief that such operations could never be decisive by themselves. It ignored the possibility that such strikes could reduce the enemy to a state of paralysis, making battles in a traditional sense unnecessary. Accordingly, it placed the role of independent armored forces in a subordinate position with regard to the actions of the army as a whole, and it held that these forces should form only a relatively small proportion of the army's total size. Thus, while the cavalry concept was a major advance over the restricted role of tanks in World War I, it fell far short of the theory of armored warfare that Fuller and Liddell Hart were advocating.

While Liddell Hart was generally pleased with the third exercise of the 1928 training season, the exercises in their entirety left him pessimistic. Their only accomplishments had been to reinforce what had been learned previously, namely, that the problem of the tactical control of a moving column had not been overcome and that traditional methods of control were far too cumbersome.[39] Collins, who had been criticized for being overly cautious during the first year, had by now become obsessed with the problem of security; in fact, on a number of occasions he even sent infantry ahead of the tanks to scout terrain.[40] Liddell Hart became so depressed by such examples of misuse that his review of the year's training hardly mentioned the work of the Armoured Force; instead, his comments were largely confined to the problems facing modernization of the army as a whole and the inadequacies that made this modernization necessary.

The official army reaction to the exercises is harder to determine, as no report was ever published. Evaluations by some of the senior commanders and members of the Army Council are available, however, and these reinforce the conclusion that a reaction had set in since the previous year.[41] One aspect of this reaction was that the 1928 tests, unlike those of 1927, were deliberately designed to hamper the Armoured Force. Burnett-Stuart admitted this openly in his report on the exercises, saying that "most of its exercises have been deliberately planned to bring out its [the Armoured Force's] limitations rather than to make a display of its powers." Montgomery-Massingberd agreed, justifying this action on the grounds that the earlier exercises had been discouraging to the older arms and that the army would never have to face a foe without a similar force of its own.[42] Since these were the men who designed the exercises, there can be no doubt as to the authenticity of this view. It is at this point, perhaps, that the full implications of the army training memorandum of the previous year made themselves felt.

Burnett-Stuart's comments on the training, which were by far the most detailed, also pointed out two basic deficiencies that were to have even more serious implications for the future: first, the lack of adequate equipment and, second, the lack of an adequate maneuver area. The multiplicity of vehicles had been a problem noted by many writers since the very inception of the force, but equipment shortages had prevented the army from reaching any solution. There was still no really effective light tank for reconnaissance, those in existence being both difficult to maneuver and very slow. The remaining vehicles of the force were an ill-assortment of tracked and wheeled vehicles—all, for practical purposes, unarmored—and coordination of movement and action between them and the tanks was extremely difficult. The tracked vehicles could generally move cross-country where the wheeled vehicles often could not. An equally serious problem was the lack of an adequate maneuvering area. Mechanized units had a radius of action far in excess of any other force in history, but their full potential could never be realized if their movements were restricted to a range of thirty or thirty-five miles. This restriction was particularly vexing because it forced them into certain patterns of movement that were highly disadvantageous to themselves and made the task of the nonarmored troops far easier than it ought to have been.[43] By themselves, neither of these matters would have been of major significance, but to an army not yet convinced of the value of independent tank operations they were another excuse for going slow.

Burnett-Stuart's and Montgomery-Massingberd's conclusions regarding the lessons of the exercises reinforce the fact that a reaction had set in against the tanks. Burnett-Stuart—whom Liddell Hart referred to as a recent convert to mechanization[44]—did not renege on any of his earlier statements, but stressed the negative lessons of the tests. This followed naturally from his admission that many of them had been designed to bring out the weaknesses of the Armoured Force, but it is clear that he was looking backward rather than forward. The main lessons of the year's training, he wrote, were that the full development of mobility had not yet been reached, that tanks were extremely sensitive to terrain, and, finally, that an independent armored force was essentially a raiding force. No one would argue with the first two points, but the third reveals that Burnett-Stuart, whatever his attitude toward the tank itself, obviously had not accepted the theory of armored warfare.[45].

Montgomery-Massingberd was even more conservative. He recognized the potential an armored strike could have against an enemy flank, but he concluded that the exercises were important primarily for demonstrating that the conventional arms could still be effective against armored forces.

He also went far toward supporting many of the claims of the cavalry: they were still extremely useful for reconnaissance, as they were quieter and less sensitive to terrain than tanks, and he recommended that they should be part of any mechanized or armored force in the future. The tank-cavalry team, he concluded, would enable each arm to make up for the deficiencies of the other and produce an all-around effective force.[46]

Perhaps even more dispiriting were the comments of the members of the Army Council, not because they were conservative, but simply because they reveal a complete lack of interest in the whole affair. Milne never bothered to write any comments at all in the file, and those of the AG, QMG, and DMO were both general and vague. In fact, all three admitted that they had not paid much attention to the tests as they were tied down with routine administrative duties.[47] Given the nature and importance of the tests plus the obvious interest the question of mechanization had aroused throughout the army, it seems clear that the minds of the Army Council were already made up: the tests had been scheduled and were so well publicized that they could not be canceled, but they could be ignored, and that is exactly what was done.

The demise of the Experimental Armoured Force occurred in November 1928 under the guise of a new series of experiments to examine other aspects of the army's mechanization program. On November 12, Milne addressed a memorandum to the Army Council in which he summarized the accomplishments of the tests to date and the decisions still facing the army. He averred that the Armoured Force had demonstrated two things: first, in the area of equipment, it not only revealed the soundness of the armored cars and medium tanks then in uses, but also highlighted the complete inadequacy of the light tanks; and, second, it revealed what he considered would be the probable future organization of an armored brigade, specifically, one medium-tank battalion and two light-tank battalions together with headquarters and supporting units. The question remained, however, as to whether or not light tanks also should be included in the standard organization of infantry brigades to increase their offensive power. To deal with this question, Milne now proposed to form a new experimental infantry brigade of three battalions of infantry and one battalion of light tanks. Because the number of light tanks was so limited, however, it was necessary to dissolve the mechanized force, but he expressed the hope that this would only be temporary, and he concluded that "during the year 1930/31 I hope to be in a positon to form the 1st Armoured Brigade as a permanent foundation of the Army."[48]

Thus ended the career of the first fully mechanized combined-arms unit in military history. It ended rather ignominiously, considering the high hopes that had accompanied its founding, and yet it cannot be considered a complete failure. It had proven that tanks could operate beyond the rigid limitations of World War I, and, more important, it convinced many people—including senior officers who had been skeptical—of that fact. In the future, the question facing the British Army would not be whether tanks had an independent role, but whether that role would dominate or be subordinate to the other arms, that is, whether the army would adopt the stategy of armored war or whether it would adhere to the strategy of attrition updated by the advent of the tank, the cavalry concept of tank warfare.

Unfortunately, the tank advocates encouraged the trend toward the cavalry concept now by arguing that armored forces should contain nothing but tanks. This stance was the product of the growing feeling that tanks could reach their full potential only if they broke loose from the traditional arms, and it was reinforced by the difficulties encountered by the EMF. The variety of vehicles had produced significant logistical as well as tactical problems so that the attraction of the all-tank concept seemed overwhelming; in one stroke, the tank advocates saw the opportunity to eliminate the two major problems facing them. Unfortunately, this attitude strengthened the belief in the army at large that armored formations were specialized units rather than all-purpose combat units and thus gave impetus to the cavalry concept.[49] The only tank advocate to oppose this view, Liddell Hart, realized that tanks needed infantry support, but even he placed their strength too low. Thus the results of the EMF were mixed. It had opened the way for two paths of development, but, by itself, gave little indication which one the army should follow.

Marking Time: 1928–31

The demise of the Experimental Armoured Force without definite plans for its replacement was naturally a blow to the tank advocates' hopes. Many of them believed it ushered in a period of reaction lasting to the mid-1930s and, in retrospect, marked the beginning of Great Britain's loss of leadership in the area of armored warfare. Liddell Hart, for example, on the occasion of Milne's retirement in 1933 reported that the CIGS's first years in office saw much more progress in army modernization than his later years.[50] And Broad and Pile, both of whom were reassigned out of the War Office in 1931 in routine changes of duty, have reported the same feelings. Broad, in fact, later recalled that "I think they [War Official officials] were rather fed up with tanks" and just as happy to see him go.[51]

There is much truth in these impressions. Certainly one of the most striking changes of this period was the virtual end to the public debate over the course of mechanization and the future of tanks that had taken place in the military journals. In the early 1920s, the *Army Quarterly* made editorial references to this question at least once a year, and by the end of the decade it was raised in every issue. The mechanization trials in 1927 and 1928 were undoubtedly the center of the army's attention. After 1930, however, the question was hardly mentioned. The *Army Quarterly* last raised it in 1930; after that, it was never raised again. Even the tank exercises in 1931 and 1934 passed by virtually unnoticed: the 1931 exercises were never covered and those in 1934 received only minor attention. It was as if the whole question had ceased to matter. No one (except Fuller, Liddell Hart, and a very few others) supported it and no one opposed it; everyone simply forgot it.

The reasons for this change in attitude are difficult to pinpoint. General Pile has expressed the view that it was because of fading memories of World War I, which led many officers to forget the lessons of that conflict and, hence, the value of tanks.[52] Another possibility is the financial pressure the army began to experience as the depression struck Britain. Throughout the 1920s, annual army expenditures had remained about £44 million—a figure only five percent of the 1918 expenditure—but in 1928 it dropped to just over £40 million. From 1931 through 1934, annual army expenditures remained slightly below £40 million and in 1932 reached a low of £35.8 million.[53] The immediate result of this was, quite naturally, to put a freeze on the purchase of any new equipment. Slight changes could be made and the army might even raise the money for the Carden-Lloyd light tanks, but these items of equipment were relatively cheap. Expensive equipment was simply out of the question, as indicated by the fate of the "Sixteen Tonner," a new medium tank built by Vickers in 1928. The design was a considerable improvement over the old Vickers medium, but its cost was prohibitive in times of financial stringency. It does seem that the army desired the tank—a few prototypes actually saw service—but the order for more had to be cut. In September 1931, Pile told Liddell Hart that "the financial stringency is awful. I think the CIGS hopes to keep mechanization where it is but I doubt if he will try and advance. Actually, though, we . . . haven't come too badly out of the crisis."[54]

Liddell Hart used his position on the *Daily Telegraph* to urge cuts in other areas of army spending to provide money for mechanization. In 1929, he noted that an increase of £18,000 for gasoline as a result of mechanization had produced a savings of £48,000 in forage for horses. The lesson was clear: mechanization would save money.[55] This became one of his favorite themes in the 1930s, and, while it was undoubtedly valid, it avoided the question of how the army could get the money for mechanization in the first place. Immediate costs in research and development and capital outlay were certain to weigh heavier than future savings to a budget-conscious government. Mechanization was not the only area of the army hurt by these drastic cuts. In the 1930 army estimates, the total for all military stores was reduced by almost thirty percent, and even the purchase of ammunition was reduced slightly, forcing the army to go into its reserves to conduct training.[56] The situation became so critical that a November 1931 War Office memorandum described the army as "barely sufficient" to fulfill its peacetime duties. The following year the chiefs of staff subcommittee reinforced this point, saying, "Except for the moral effect of its [expeditionary force's] presence on the Continent it could have little effect on the future of the campaign."[57]

These financial restrictions had a serious effect on the mechanization and, indeed, on the whole outlook of the army. Directly, of course, they slowed the process of mechanization. Indirectly, they had an even more serious effect. The cuts placed the army in a tenuous position in which every unit and officer undoubtedly felt extremely insecure. This insecurity, in turn, bred a generally defensive attitude in which change was seen not as progress, but as a potential threat to one's own position. The government reinforced this attitude by opposing cuts in army manpower, as they would add to the already severe unemployment problem. In short, the cuts in army expenditures bred fear and fear bred stagnation.

Another consideration, more difficult to assess precisely but nonetheless important, was the change of attitude Great Britain as a whole experienced at this time. The two standard histories of this period—written by A. J. P. Taylor and C. L. Mowat—both make reference to a loss of spirit in England in the late 1920s, Taylor referring to it as "Half Time" and Mowat, as "Sitting It Out."[58] It was a period of confusion and uncertainty, and the British people looked for claiming influences rather than heroic leaders who promised drastic remedies. In the words of Robert Graves and Alan Hodge:

> A characteristic of the thirties in England was an attempt to be reasonable about the confusion into which the new theories of physics, astronomy, sex, and economics had plunged thinking people. . . . Inside there was a general consensus of opinion: Never to do what the Russians had done and the Germans and Italians were doing—pull the house down and build up from new foundations—but to continue patching and revetting and bracing so long as it would stand. The country was still sound at heart.[59]

That the British Army could have remained isolated from this general atmosphere is improbable, and there is considerable evidence to the contrary. The disappearance of the controversy from the military journals, for example, was undoubtedly in part the result of the growing belief that the basic questions regarding mechanization and the future of tanks were resolved. By 1929, the army was coming to accept the cavalry concept of tank warfare. As in Great Britain as a whole, there would be no attempt "to pull the house down and build up from new foundations," as Fuller and Liddell Hart were demanding; rather, there would only be a gradual process of improving the power and efficiency of a military machine that they continued to believe was founded on sound doctrines and a correct understanding of the principles of war. As C. L. Mowat noted, "The mood of the early thirties was not heroic."[60]

The army, however, expressed little dissatisfaction with the progress being made in mechanization. Rather, there was a sense of self-satisfaction in many quarters. The armored force was, after all, the first of its kind in the world and a new small tank, to become famous as the Carden-Lloyd Mark VI, was going into production and making a strong impression in military circles abroad. Moreover, besides continuing to produce new vehicles for the Royal Tank Corps, the British Army had also begun to replace horses with vehicles in many other units, although at an extremely slow rate. The cavalry machine-gun squadrons were completely mechanized, and a similar process was moving ahead in the infantry. The artillery was also being mechanized slowly, and in April 1928 the first British cavalry regiment replaced its horses with vehicles, the Eleventh Hussars being converted to an armored-car unit, reportedly with much sorrow among its members but without apparent protest.[61] Such changes gave substance to the army's view that it was modernizing as rapidly as government financial restrictions permitted and in April 1929, the *Army Quarterly* smugly told its readers:

> Unlike our military authorities, the American experts seem to be by no means wholly converted to the mechanization of their forces. The troops and the infantryman are still looked upon as the backbone of the Army, and it is interesting to note that in the American Army there is only one armoured car unit in being at present.[62]

But this was little comfort to the tank advocates, who held that equipment changes in the traditional arms were essentially cosmetic. As long as the army adhered to its strategic doctrines, however, such criticism would fall on deaf ears.

There was one event at this time, however, in which the tank advocates took a great deal of pride: the publication of the first official manual on armored warfare. Entitled *Mechanized and Armoured Formations,* it was written by Charles Broad assisted by George Lindsay at the request of Milne, who told Broad to write out what he considered to be the lessons of the 1927 and 1928 excercises. The "Purple Primer," as it came to be known, was significant primarily because it marked the first time the theories of Fuller and Liddell Hart were applied to the realities of maneuvering and controlling armored fighting vehicles on the ground. Broad was an outspoken advocate of fast-moving armored warfare, but the tone of the "Purple Primer" was consciously sober, for he, unlike Fuller, was more interested in gaining converts than in "making points." Yet the message was clear. The manual began with a brief but complete description of the theory of armored warfare as it had developed to that point. It stressed that the mobility and endurance of armored formations were far greater than those of infantry, and asserted that movements of thirty to fifty miles a day for as long as six days could be expected from them. Such mobility, it continued, could have a psychological as well as a material effect on an enemy unsuitably equipped to meet them. He could become uncertain and confused, perhaps even paralyzed. As a result, the armored-force commander could concentrate a superiority of force at the decisive point, or, if the enemy's state of mind had deteriorated sufficiently, possibly achieve a decisive success with numerically inferior forces.

By far the most important section of the book was that dealing with the organization and control of an armored force. Thus far, the organization of such a force had been an admittedly offhanded affair. Reference has already been made to the problem of the multiplicity of vehicles, and this continued to plague the Royal Tank Corps. Broad's solution to this problem was to eliminate everything but tanks from an armored force. It would contain only light tanks for reconnaissance and medium tanks for striking power. These would be organized into separate battalions and grouped together into brigades comprising two battalions of light and one battalion of medium tanks. The problem of control was far more difficult because tanks operated in a more dispersed pattern and moved more rapidly than the other combat arms. Broad's innovative solution lay in the formulation of a new type of battle drill. The battle drill (which is analogous to the more familiar parade drill except that the formations and movements are based solely on tactical requirements) was already in use in the infantry, but its transformation to armored units was no simple matter. Broad's system was designed to overcome these problems and, in fact, turn them into advantages. He established basic formations and patterns of movement that were simple and yet extremely flexible and capable of rapid execution so that the mobility of the armored force was emphasized to the utmost. In essence, these formations consisted of a forward element that conducted reconnaissance and pinned the enemy to his position once he was located, and a shock element that maneuvered around an enemy so fixed and struck him from a unexpected quarter. The formations were geometric—mostly squares or triangles—and were designed to be altered rapidly to meet new tactical or geographic conditions. The system was capped by a series of flag signals designed to enable the commander to control the movement of the tanks with parade-ground precision. As he himself would demonstrate two years later, in the hands of an experienced commander, this system was ideally suited to the requirements of rapid armored warfare.[63]

The "Purple Primer" was valuable because it established a model for further thought and discussion. It was made available throughout the army and to the public at large, an indication that at least some people in the War Office wanted to see the discussion on mechanization continued. This impression is reinforced by a lecture given at the War Office on March 26,

1929, by Major General Bonham Carter. The general, who was not associated with the tank advocates, was notably cautious about the pace of mechanization, but he treated Broad's work with considerable respect. He emphasized that tanks, their organization, and their tactics were still in the formative stages of development and that the future remained uncertain. In this light, Bonham Carter presented the new tank manual as a work designed to stimulate further thought as well as to summarize developments thus far and he urged his listeners to study these problems with unbiased minds.[64]

For the time being, however, the army diffused its efforts to develop an effective tank force. Training for 1929 concentrated on three series of test exercises, each of which involved mechanization in one form or another: the first involved the experimental infantry brigade and was designed to test the value of adding a light-tank battalion to its organization; the second involved the cavalry and was designed to test the value of armored cars in cooperation with horsed units; and the third was an exercise involving tanks only and was designed to find the right proportion of light tanks to medium tanks in the tank battalion.[65] All of the exercises were thus organizational rather than tactical in nature.

The tests ran into a serious problem before they even began, however, and the army's solution sheds an interesting light on its attitudes at this time. In order to conduct all of the tests, the army had to secure 212 Carden-Lloyd Mark VI light tanks. These were modified versions of the earlier light tanks, which had not proven successful in 1927, and the new vehicles were considered necessary if the tests were to effective. This presented a problem because routine procurement procedures precluded having the tanks in time for summer training. On November 28, 1928, the secretary of state for war asked the chancellor of the exchequer whether the army could place an order for the required machines—costing a total of £112,360—to be paid for from the following year's budget. While the machines would not have to be paid for until that time, the secretary explained that, if the army had to wait until the 1929 army estimates were approved in March to order them, they would not arrive in time for that summer's training. He concluded:

> The programme stands at the head of our mechanization schemes and will be given absolute priority within whatever total is fixed when we come to consider the Estimate. Whatever else has to suffer, I think we should be shortsighted to miss this chance of gaining indispensable experience.[66]

The army's willingness to take a step such as this is further demonstration that its supposed conservation on mechanization was not nearly so pronounced as often assumed. But as most of these tanks were to used for the test involving infantry and mechanized cavalry, it also indicates that the army was still primarily concerned with increasing the combat power of the traditional arms and that it had made no movement away from the strategy of attrition.

The 1929 training season was largely restricted to small-scale exercises, the results of which were of marginal importance to the armored forces. From the infantry-brigade tests, the army concluded that, while the light tanks provided added power and mobility to the unit, such an arrangement unduly restricted the tanks' mobility and was, on balance, unwise. From the tank exercises the army concluded that light and medium tanks should operate in closest cooperation, and it became accepted that an organization of mixed battalions of light and medium tanks was superior to one of separate light- and medium-tank battalions. The cavalry tests demonstrated that partial mechanization would further increase that branch's mobility and power, but most observers agreed that the value of the horse was still evident in its ability to conduct reconnaissance in certain types of terrain.[67]

If the tests themselves lacked drama, there were, nonetheless, indications that the army's general attitude toward mechanization and tanks was becoming more positive. Liddell Hart, who, after expressing disappointment that the 1929 exercises were on such a small scale, reported that "the old antipathy to 'new-fangled' things in general, and to tanks in particular, is fading." He added that an increasing number of officers, especially generals, were coming to see the value of tanks and even to look for their support whenever possible. The major problem still facing the army was the absence of an adequate number of tanks because of the failure of many officers to get rid of the "man-power obsession"—that is, the belief that the strength of the army depended solely on the number of men in the ranks.[68] He felt there were indications, however, that this, too, was subsiding. This analysis was given substance by a 1929 *Army Quarterly* editorial—a journal that continued to advocate a cautious approach to mechanization—which declared that

> as the process of mechanization continues, it is likely that the strength of armies will be measured in the same way as that of navies, and the military power of a country may depend not so much on its manpower as on the quality and quantity of its mechanical equipment.[69]

This increasing emphasis on the importance of tanks was reflected even more dramatically in the 1929 edition of the *Field Service Regulations.* Although the two earlier postwar editions of the *Regulations* had described tanks as a vital instrument of modern war and had stressed that victory could only be achieved through the combined efforts of all arms, they had explicitly subordinated the role of tanks to that of infantry. Indeed, they had asserted that "infantry is the arm which in the end wins battles. The co-operation of the other arms is necessary, but neither separately nor together can they defeat the enemy."[70] In the new edition, however, cooperation among the arms was depicted in more balanced terms, and the importance of armor correspondingly was increased. "The proper co-operation of all arms wins battles," it stated, "and enables the infantry to confirm victory" (rather than achieve it). It continued:

> Infantry is assisted mainly by artillery, and on occasions by tanks, in its task of closing with the enemy. Without mobile troops and aircraft, the other arms cannot move in security and are unable to gather the fruits of victory, while the cavalry is at a great disadvantage unless accompanied by artillery and armoured vehicles.[71]

The *Regulations* also described the limitations of infantry in far more forceful terms than ever before, which in effect placed that arm in a subordinate position in offensive operations. Explaining that because of its vulnerability advancing infantry "requires . . . superiority of fire" in order to succeed, it admitted that "it is only in exceptionally favourable circumstances of numbers, morale and ground that infantry in the attack can produce superiority of fire unaided." To succeed, it continued, the support of artillery and "armoured fighting vehicles are the best methods" available. Emphasizing the same point from the opposite perspective, the *Regulations* stated that "infantry acting defensively has little to fear from hostile infantry attack which is unaided by other arms."[72]

Moreover, the independent role of tanks received far greater emphasis than it had in the *Regulations* of five years earlier. In that edition there was only one reference to such operations, and that was placed in a rather inconspicuous position at the end of the section describing the offensive role of tanks. In the new edition, there were no fewer than three such references, and they were given far greater prominence than before.[73] Correspondingly, the role of cavalry was somewhat more limited. It still was described as having "the power to deliver a mounted attack," but the description of its duties revealed a growing awareness of its limitations. In laying the principle for independent mobile operations, for example, it stated: "Highly mobile forces of several arms may be expected to play an important part in future wars, but cavalry unaided will have little striking power, except in certain theatres where the enemy's armament and fighting value are relatively inferior."[74]

The growing awareness of the power of armored formations did not signal a shift away from the strategy of attrition toward the strategy of armored war, however; the Napoleonic model of war still reigned supreme, although there were modifications of some of its elements. Most notably, there was a greater realization that war consisted of more than great battles. In the 1920 edition, it will be recalled, the *Regulations* had described the importance of battle in unequivocal terms.[75] In the 1929 edition, on the other hand, while battle was still described as "the decisive factor in war," nonmilitary considerations were given much greater emphasis. War was described as a contest of wills in which "diplomacy, economic influence applied in the form of financial and commercial restrictions, and—in the last resort—the use of armed forces" may all be applied. Within this last sphere, the role of the armed forces was "to break down the resistance of the enemy's armed forces in furtherance of the approved plan of campaign."[76] This description paralleled the ideas put forward by Fuller and Liddell Hart—although there is no indication that they had any direct bearing on these changes—but was not matched by any change in the army's concept of military operations. Indeed, the *Regulations* did not change a single word in its general description of the principles of military operations quoted earlier.[77]

Further recognition of the increasing role of armor came the following year with a new revision of the contingency plans for an expenditionary force. In 1923 and 1924, such plans had been drawn up and included a tank battalion for each of the infantry divisions for direct support. In this latest revision, however, an additional "armoured force" was assigned to general-headquarters troops as part of the first contingent to be sent abroad. The exact composition of this force was " to be decided later," and its role was not specified; but it is clear that such a force was designed for independent operations rather than to provide additional direct support for the infantry. Given its place in relation to the expeditionary force as a whole and the fact that the army was hoping to form a tank brigade at this time, it was a clear sign that the army was putting the expanded role of armor described in the 1929 *Regulations* into practice.[78]

Yet both the 1929 *Regulations* and this revision in the composition of the expeditionary force only reinforce the fact that the army's growing awareness of the power of armor was still restricted to the framework of the strategy of attrition, and that it had not moved any closer to the theory of armored warfare. The "armoured force" was merely a raiding force; useful, but scarcely capable of delivering the paralyzing blow that Fuller and Liddell Hart perceived to be their true role.

Unfortunately, the 1930 training season did little to sustain the hopes raised by these changes. Milne's hope that a tank brigade could be permanently formed that year were quashed by the financial restrictions of the government,[79] but a temporary brigade of the Second and Fifth Battalions of the Royal Tank Corps was formed and employed in an interdivisional exercise between the First and Second Divisions. Their use, however, reflected little imagination. Liddell Hart's optimism concerning the growing acceptance of tanks was replaced by a comment on the "de-mechanization of its [the Army's] mind," and he was strongly critical of the commanders of the divisions for failing to make the maximum use of the training area available for the maneuvers.[80]

The First Brigade, Royal Tank Corps and After, 1931–34

The year 1931 brought about a resurrection of hopes for the tank advocates and important advances in the development of tank tactics. The occasion was the formation of the First Brigade, Royal Tank Corps, on April 1, 1931, under the command of Brigadier Charles Broad. The brigade was composed of the Second, Third, and Fifth Battalions, Royal Tank Corps, each composed of a section of close-support tanks and three mixed companies of five Vickers medium tanks and seven light tanks. In addition, a further light-tank battalion was organized on a skeleton basis and attached to the Second Battalion, giving the brigade a total strength of eighty-five medium tanks and ninety-five light tanks.[81] The unit thus reflected the lessons of the tank exercises that had been held since 1927: it was all tank and it was so organized as to provide for the closest cooperation between the medium and light tanks. It did, however, have serious weaknesses. The most obvious to observers was the virtual absence of any true light tank. In 1930, a new light tank, the Mark II, had been adoped by the army, but financial problems had limited the army's orders to only ten machines by the time of the exercises. For the remainder, the Carden-Lloyd machine-gun carriers still had to suffice. Even more important than the weaknesses in equipment was the fact that the brigade was a temporary unit from beginning to end. Broad was between assignments during the period—he was never taken off half-pay during the exercises—and even had trouble getting an office and furniture for his headquarters. Furthermore, the brigade actually had only four weeks to train as a unit, three of which had to be spent in battalion-level training. In Broad's words, it was a "half-baked affair."[82] It was also a tribute to his superior organizational and training ability that anything at all was accomplished.

The purpose of the tests was basically to put the theories Broad had expounded in the "Purple Primer" into practice, in short, to show that tanks could be maneuvered tactically in masses. This had not been done effectively before. Tanks had been shown capable of making sweeps around an enemy flank as early as the 1925 army maneuvers, but each time, the actual attack itself had been poorly executed. It was Broad's job to correct this, and to do it had had an invaluable new aid in the form of primitive radio sets. Prior to the exercises, he had spoken with a signal officer by the name of Fuller[83] about the problem of tank control and Fuller provided enough crystal sets for the exercise. They were primitive devices and difficult to hear over the noise of the tanks, but they sufficed for the purpose.

Broad went beyond the restrictions of his orders, however. He was not merely a tank man, but a strong believer in the concept of fast-moving armored warfare, and he used this opportunity to show not only that tanks could be maneuvered tactically, but also that they were actually capable of conducting the type of operations Fuller and Liddell Hart had been advocating for a decade. Surprising as it may seem, one of the first problems Broad encountered in this was the need to persuade many of the tank men in the RTC itself that this was possible. They had become so accustomed to working in the role of infantry support that many had yet to be convinced that they were truly capable of doing more. Broad's approach consisted of a series of lectures he gave to each of the tank battalions at their home stations before the brigade came together. Significantly, he lectured not just to the officers, but to all ranks so that the spirit of the new ideas would permeate throughout his command. The results of the exercises demonstrated his effectiveness, although he has credited much of his success to the fact that the Royal Tank Corps recruited "a highly intelligent class of men."[84]

The actual training of the brigade again reflected Broad's careful and well-thought-out methods. He devised a series of tactical situations and had each battalion run through the course. Normally, when the battalion came under fire, the light tanks were sent forward to pin down the enemy while the medium tanks maneuvered into position for the kill. Each battalion practiced the exercise until it gained the desired proficiency and then was sent on to the next one. After three weeks of this, the brigade came together for the final series of exercises, which followed the same format as the battalion exercises. Broad made effective use of the crystal sets and on a number of occasions established his own position a mile or so away from the rest of the tanks as a dramatic example of the type of control that he could exercise. Reporting in the *Daily Telegraph,* Liddell Hart was ecstatic about the results. This was the type of tank warfare he had been urging the army to adopt, and he had no doubt that the results justified his predictions. Under headlines such as "Revolutionary New Tactics," "Infantry in Eclipse," and "Tanks Irresistible in Attack," he told the story of Broad's maneuvers and heralded the dawn of a new era in warfare.[85]

The final exercise of the First Brigade took place on September 17. It consisted of an attack by the tank brigade against a moving column of infantry. The tanks were to approach in a basic triangle formation—one battalion forward, two back—with Broad in the lead, and, when the infantry was spotted, the lead battalion was to move forward and pin it down while the other two moved around the flank to strike from the rear. The concept was by no means new; its uniqueness lay in the fact that it was being carried out by tanks moving rapidly in mass. This final exercise was especially significant in that it received the first and only real recognition of the brigade's existence in the form of a visit by a number of general officers, including the entire Army Council.

The exercise worked exactly according to plan. The tanks maneuvered with a precision and assurance that could only have come from intensive training. The whole operation lasted only slightly more than an hour—despite the fact that the maneuvering battalions had to cover a total of ten miles in their flanking march—and the infantry was swept from the field. Liddell Hart reported:

> Of those who saw it for the first time, that warfare must have had a rather bewildering impression—as bewildering, doubtless, as the small and quick changing maneuvers of the Parthian horse seemed to be to the solid legionnaires of Crassus.[86]

Indeed, the exercise worked too well. To the observing generals, accustomed to a far slower pace of movement, Broad's exercise could only have seemed a carefully rehearsed scheme. They concluded that he must have spent the entire time rehearsing this one exercise over the same ground just as a band practices its complicated marches. The idea that this could have been done in any type of tactical situation was beyond their comprehension. Milne was apparently furious, feeling that Broad had tricked him and the whole exercise was dismissed. Broad recalled:

> After the exercise I sent in my report and then went on to my next job. I never heard another word about it. . . . That concluded my personal work with tanks as I was never employed in a job with them again.[87]

Unlike the exercises of 1927 and 1928, those of 1931 excited no interest in the army as a whole. In a summary of the training year, an article in the *Journal of the Royal United Service Institution* declared that "during the past training season, no exercises of a spectacular nature were undertaken."[88] It did list, however, a number of events of some interest, including an artillery exercise using live ammunition, a number of staff exercises and, finally, the exercises of the tank brigade. The author praised the state of training the brigade was able to achieve in so short a period of time, but he concluded that their importance was not great, as they were "rather like a drill test than an operation required often in war."[89] Thus were dismissed the most significant exercises held by the British Army since the end World War I.

The dissolution of the tank brigade was naturally a great disappointment to the tank advocates, but their spirits were maintained by the hope that it would be resurrected the following year. This was, indeed, the case. In August 1932 the brigade was reconstituted for the summer's training under the command of Brigadier K. M. Laird. It was larger than the brigade of the previous year, being composed of 231 tanks of all types, and had fifty new light tanks that were decidely superior to the Carden-Lloyds they had replaced. There were still not enough of the new tanks to go around, however, and the Carden-Lloyds were still in use as were the Vickers mediums, now approaching their tenth year of service.[90]

The 1932 exercises differed from those of the previous year in that they included tank-versus-tank actions and that the tanks maneuvered over far greater distances. In both instances, however, it was clear that they were building on the experience of Broad's work. This was especially true of the battle drill, which was again widely used without any difficulty as the personnel of the brigade were essentially the same.[91] The tank-versus-tank battles pitted the Second and Fifth Battalions, Royal Tank Corps, against each other in a series of short exercises. Both battalions were led by experienced tank officers and, as described by Liddell Hart, were fast-paced affairs reflecting both the experience and skill of the commanders. On September 13, the brigade came together for a series of exercises as a unit. The most impressive of these began the following day, when they undertook an operation designed to penetrate deep into enemy territory and disrupt his communications. It began with a thirty-four-mile march to the point of departure—accomplished in only two and a half hours—and was followed the next day by a wide flanking move that covered an additional fifty miles before dusk. It was an impressive demonstration of mobility and, given the fire power of the brigade, showed perhaps more than anything thus far the strategic power of the new arm.

Unfortunately, the overall impact of the exercises was slight. As in 1931, the "enemy" was imaginary, and the exercises themselves attracted little attention apart from Liddell Hart's articles in the *Daily Telegraph.* Most senior officers seem to have felt that, when the army came up against "real" troops, such movements would be impossible. When the tanks did work with the infantry, it was still as an infantry-support weapon, which was most unfortunate for both arms. The infantry failed to see the potential of armored mobility and the tankers became increasingly discouraged about cooperating with the infantry because of the fetters such cooperation entailed. As a result, each arm continued to go its own way to the detriment of both.

On a more progressive note, October 1932 saw a little-known but highly significant document issued by the War Office entitled "Report of the Committee on the Lessons of the Great War." It was prepared by a select group of senior officers appointed to discern the lessons of the war and the extent to which they were then being applied in the British Army.[92] Although it did not deal directly with any of the major controversial aspects of the conduct of the war, these issues obviously could not have been completely avoided, nor, indeed, were they. In fact, by implication, if not by direct statement, the committee strongly condemned many aspects of the conduct of the war on the Western Front. Its analysis of the massive offensives on the Western Front, for example, did not criticize the British high command directly, but its conclusions were hardly favorable to those responsible for planning and conducting those offensives. It asserted the key to success lay in surprise (which virtually all of the great offensives of 1914–17 had ignored) and declared that neither the addition of more artillery nor of more infantry in themselves would have altered the chances of success. The artillery could be countered by a withdrawal of enemy troops beyond its range of fire, and the addition of more infantry would only result in adding to the casualty lists. In the words of the report, "An operation which has once failed will not succeed without the introduction of a new factor, and that factor is seldom more infantry."[93] It further stated:

> The greatest lesson to be learned from the world war is that no attack in modern war is feasible or likely to succeed against an enemy in position unless his resisting power has already been paralyzed by:
> *(a)* some form of surprise; or
> *(b)* preponderating fire, powerful enough to produce the effect of surprise.[94]

The army's training manuals, it concluded should, "emphasize more strongly the vital importance of surprise and of the indirect approach."[95]

Having, indirectly at least, condemned the conduct of the offensives, the report ignored the larger question of their effect on the outcome of the war. It did, however, discuss one of the solutions to that stalemate, namely, the tank, and in so doing gave another indication of the army's growing awareness that tanks would play a powerful role in any future conflict. The report stated that

> the performance of the tank on the Western Front and even more what they could obviously have done in other theatres place their value beyond a shadow of a doubt.[96]

Given a repetition of these circumstances, moreover, it recommended the appointment of younger commanders who would provide more vigorous leadership, the placement of headquarters closer to the front and

> lastly comes the suggestion that the whole plan of battle might be simplified by greater reliance on armoured fighting vehicles, their direction and control being assisted by aircraft communications.[97]

The other combat arms were still necessary but their role would be "to secure successive bases from which tanks can make fresh bounds."[98] Clearly, the punch of any offensive was the tank; the other arms were merely supporting forces or forces of occupation. This was further emphasized when it declared that "the chief lesson [of the war] is the necessity of increased covering fire to make movement over a bullet and shell swept area possible." To achieve this, "it seems we should not be wrong in expanding our resources in tanks to clear the way for the infantry," and it concluded:

> To sum up, the problem is to increase the proportion of tanks certainly, artillery probably and other covering weapons possibly, and this without increasing the size of the division, which would unfit it for the "average condition."[99]

The significance of this report lay in what it both did and did not say. In the first place, its emphasis on further tank development was another indication of the army's awareness that tanks would not simply be nice to have in future war but would be essential. In visualizing the use of tanks, the committee was still wedded to the idea of the infantry tank, but not to the slow pace of a traditional infantry advance; rather, it urged that the infantry should be motorized and follow the tanks' advance as quickly as possible. Furthermore, the report strongly emphasized the importance of flexibility, surprise, and the indirect approach whenever possible. It did not, however, reach the tank advocates' ultimate conclusion that the object of the armored threat should be the "brain" of the enemy army. It completely ignored both the concept of deep armor penetrations and the principle of strategic paralysis. Thus, while the report was the most far-reaching criticism of the army's conduct of the World War, it failed to accept the full measure of the tank advocates' criticism and, by implication, remained wedded to the strategy of attrition, however more efficiently it wished to see that strategy applied.

In February 1933, Field Marshal Sir George Milne retired as CIGS. He had occupied the post since 1926 and had been given two extensions beyond his original four-year term so that he might continue his work in the mechanization of the army. Despite these extensions and the many progressive statements he had made on the subject, however, his achievement was limited. The Royal Tank Corps still had only four battalions and these had yet to achieve full strength, although they were in far better shape than in 1926. As for the remainder of the army, only a few units had been mechanized, so few, in fact, that Liddell Hart wrote on this occasion, "For practical purposes, 'mechanization' is a catch word rather than a reality." Among the reasons for this lack of progress, Liddell Hart noted the government's financial stringency, varying imperial conditions, vested interests, the doubts of conservatives, and underlying all of these, Milne's own character. In Liddell Hart's view, the retiring CIGS was, despite his progressive statements, essentially a cautious man and one too uncertain of his own views to override others who urged even more caution. Furthermore, he felt Milne was also suspicious of strong personalities and tended to keep shy of them.[100] He was a complex man, but whatever the reasons, his tenure as CIGS saw little real change in the British Army despite the initial hopes he had raised.

Milne's successor was General Sir Archibald Montgomery-Massingberd. He had been born in 1871 in Ulster—and was thus a member of that Anglo-Irish gentry, which had produced so many officers—and, like his predecessor, had entered the artillery following graduation from Woolwich. His career was primarily that of a staff officer, and, although he played a key role in planning the battle of Amiens when serving as chief of staff to the Fifth Army, he had never held a major wartime command. As a result, he was so little known in the army at large that when he became CIGS, the *Army Quarterly* commented that "General Montgomery-Massingberd is no longer a young man and he has really yet to show what there is in him; his appointment seems to be very much an experiment—an experiment which the Army will watch with interest."[101]

Montgomery-Massingberd's attitude toward mechanization remains clouded in controversy. Socially, he was a product of the country gentry, and his brief memoir, written just before his death, reveals a man to whom the horse and the hunt were always very dear.[102] On a number of occasions he expressed the opinion that horses were still valuable in modern war and that mechanization entailed many difficulties that required it be undertaken slowly and carefully. As a result, he has sometimes been pictured as an archconservative to whom mechanization was complete anathema.[103] On deeper examination, however, this view appears to be an oversimplification. It is undoubtedly true that he strongly disagreed with the theory of armored war as expounded by Fuller and Liddell Hart. But it is equally true that he saw the need for and value of tanks, as indicated by his formation of the First Tank Brigade on a permanent basis in 1934 and by his continuous efforts to create the Mobile Division (later renamed the First Armoured Division) afterwards. Indeed, when Hobart began his work with the Tank Brigade in 1934, he wrote Liddell Hart that "most of the great ones scoff, of course, but we are lucky indeed in at least having so far-seeing, resolute and open minded a CIGS who is giving us a chance to try and is so remarkably understanding."[104] From these conflicting views comes the picture of a man who was a curious blend of the old and the new. Montgomery-Massingberd deeply cherished army traditions and felt that they were extremely important, but he also recognized that mechanization was necessary if the army was to remain an effective fighting force and he was determined to pursue this course to the best of his ability. His methodical nature determined that this would be a slow process, but given the government and popular attitudes of the mid-1930s, it was an approach fully in keeping with the times and the only realistic course he could have followed. Unfortunately, his foresight in many areas was marred by the fact that he did not envisage the use of tanks beyond the cavalry concept.

Montgomery-Massingberd's first training season as CIGS was largely uneventful. Liddell Hart commented at the time that "the general verdict among officers who watched the Army Exercises of 1933 seems to be that it was one of the dullest years since the war."[105] The training was largely concerned with guerrilla operations, and while it did allow for a much wider scope for independent action among junior officers than was customary, it afforded no real experience for higher commanders or the tank units. The role of tanks, in fact, was wholly restricted to that of infantry support.[106]

The First Tank Brigade, 1934

The 1933 training season was scarcely over when planning began for much more significant tank exercises during the following year. In November, the First Tank Brigade was formed on a permanent basis—a step Milne had hoped to be able to take as early as 1930 but had been prohibited from doing by financial considerations—and P. C. S. Hobart was appointed its commander.[107] The choice of Hobart proved a masterstroke (albeit not necessarily intended), for he was not only a strong-willed, energetic commander, but he had also commanded the Second Battalion, RTC, under Broad in 1931. He thus combined a thorough understanding of the theory of armored warfare with experience and a determination to see that the First Tank Brigade was transformed in his image. Within weeks of his new appointment, he met with Lindsay and Broad to discuss the "character" of the new unit. As leading advocates of the strategy of armored warfare within the army, they naturally believed the brigade should be employed solely on independent operations and they sought to have the new CIGS issue a directive to this effect.[108] They achieved a limited success toward this goal when, on January 24, the War Office issued a detailed study of the Tank Brigade that stressed its independent role. It stated that missions such as disruptive raids, flank attacks in conjunction with assaults by the main body of the army, and pursuits offered the greatest possibilities for the brigade. Correspondingly, it firmly downgraded direct support operations, declaring, "Against a reputable and well-equipped enemy, this method will rarely be justified."[109] This was a long way from the comments the War Office made following the 1925 army maneuvers, but it still only reflected acceptance of the cavalry concept of tank warfare.

During the spring of 1934, preparations steadily went forward for the season's training. From May 8 to 12, a Royal Tank Corps staff exercise was held under Hobart's direction at Cambridge to study the problems of tank operations deep in the enemy rear. The exercise was not framed to determine the feasibility of such operations but to develop techniques that could be used in them. In this light, the report on the exercise stressed that the brigade should move over a wide front to keep the enemy confused; that it should threaten alternate objectives; and that, while the threat of tanks might have a great effect on an enemy, such a threat must be accompanied by the actual destruction of bases, which meant that medium tanks were an absolute necessity. It further stated that "the cooperation of the R. A. F. at all stages appears to be a factor vital to success" for the purposes of reconnaissance, protection from enemy air attacks, aerial resupply, and aerial bombardment to replace artillery, which would probably be out of range.[110]

In order to accomplish this, however, improvements would have to be made in equipment and training. Unfortunately, this occurred only very slowly. The brigade did receive a number of new light tanks during the spring, but most of its equipment still consisted of old Vickers medium tanks and Carden-Lloyd tankettes, both totally inadequate for the tasks it faced. The report on the Cambridge exercise had emphasized that logistical problems represented the greatest immediate threat to the brigade's success and recommended that a five-ton truck and a 500- to 1,000-gallon gas truck would be of great help, but little could be done in the short time left before training began. As a result, the brigade had to rely on a confusing array of vehicles throughout its first year of training.[111] Furthermore, although the formation of the brigade had been announced in November 1933, it did not come into being officially until April 1, and the units were not actually brought together until June. As a result, Hobart was particularly pessimistic about the possibilities of the summer and wrote Liddell Hart, "Do not expect a high degree of efficiency."[112] This was one of the few errors in judgment he would make that summer.

The First Tank Brigade differed somewhat from its predecessors in organization, for it reflected not only the lessons of earlier exercises, but also the fact that experimentation was still going on. It consisted of four battalions, three mixed and one, the new addition, light. Each of the mixed battalions—now designated "medium"—consisted of a headquarters section of four medium tanks, and three mixed companies composed of a command tank, a section of seven light tanks, a section of five medium tanks, and a section of two close-support tanks (which were actually medium tanks). The light battalion consisted of a headquarters section and three companies of varying organizations.[113] The focus of the brigade's first training season was to study "the strategic or semi-independent role of not less than a tank brigade against some rearward enemy objective."[114]

After about a month of gunnery and battalion drill, the brigade assembled on August 25 to begin a series of exercises to test its mobility and control techniques. The first exercise was "close-order drill" controlled by radio and was followed by brigade maneuvers over open country. In the latter exercise the brigade spread out over a front of about ten miles and a depth of about five miles ahead of an imaginary main force, with the medium battalions moving over two or three roads on a front of from three to five miles. Although this technique obviously added to the problems of control, it also made it far more difficult to detect the brigade from the air and to determine its objective.

Following this, a series of tactical exercises were held in which the brigade was given various missions to accomplish in the face of conventional infantry and cavalry units. These included operations over a wide area behind enemy lines, maneuvers against and attacks on a large city, and an attack on an enemy truck convoy on the march. The tanks took on the conventional units with great relish and decimated them so badly that one of the umpires was approached by officers from these units who "complained that the troops of the Aldershot Command [which were pitted against the tanks] were being put up as a cockshy for the Mobile Force and being made fools of."[115] The tank advocates, however, were ecstatic. Liddell Hart reported that the exercises "should prove beyond a cavil the possibility of a tank force in making an effective move over a hundred miles in the twenty-four hours."[116] The Tank Corps was riding the crest of a wave.

The ride did not last for long. In September the Tank Brigade was combined with the Seventh Infantry Brigade (mechanized), a mechanized field-artillery brigade, and other supporting units to form what was, in effect, an armored division, though it was never so designated. For a final exercise it was ordered to conduct a deep raid behind enemy lines to destroy supply and command centers in the face of a conventional force composed of an infantry division, a cavalry brigade, and two armored-car companies. In theory this appeared to be an ideal mission for such a force, but military exercises can be like statistics: present them the right way and they can prove anything. In this case, the conditions for the exercise were set by General Sir John Burnett-Stuart, then general officer commanding, Southern Command, who felt the earlier exercises had been weighted too much in favor of the mobile arm and that a corrective action was necessary.[117]

The basic design of the exercise was simple.[118] The mobile force was given the mission of penetrating some sixty miles behind enemy lines and attacking an enemy supply and command complex spread out over 140 square miles. They were given three days—September 19–22—to accomplish the mission, withdraw thirty miles to safety, and prepare to support an attack by the main force of the army. As the mobile force was completely mechanized and its opponents had no mobile units except two armored-car companies, one immediately is inclined to agree with an observer who noted on the eve of the exercise, "Like the majority [of the people there] I wondered what on earth the 1st Division was going to do to protect its immobile and vulnerable self."[119] Beneath the surface, however, severe restrictions had been placed on the operations of the mobile force. In the first place, it was given no time to train together. The commander of the force, Major General Lindsay, had barely a week to organize his staff, and, although he had long been one of the leading tank advocates, he had never commanded even a tank brigade let alone a combined force of this size. Even more important, however, were the restrictions placed on him in the conduct of the operation. He was not allowed to leave the starting point until 2:00 A.M. on September 19, giving him only four hours in which to move before daylight on the first day, and his approach routes were virtually restricted by Burnett-Stuart to only one avenue. Finally, the target area was relatively concentrated: 140 square miles might seem like a considerable area, but it measures barely twelve miles by twelve miles, by no means a difficult area for an infantry division and cavalry brigade to defend even if they lacked modern means of mobility. As Liddell Hart later pointed out, to assume that there would be only one profitable target to attack for sixty miles behind an enemy's front lines was totally unrealistic in modern war; rather, there would be many profitable targets spread out over a far greater distance and much more difficult to defend by conventional troops.[120]

The exercise began with a move of forty miles by the infantry brigade to secure a bridgehead across a river barrier and was accomplished by daylight on September 19. Lindsay's plan was to seek cover during the daylight hours and move up the tank brigade during the second night, in preparation for the final move to the objective during the third night, and attack at dawn the following day. Unfortunately, the advance force was discovered—apparently because the area was visited by War Office officials and others "until the approaches looked like Piccadilly at 12:30 P.M. in May"[121]—and it was subjected to heavy bombardment from the air. Still, the tanks were able to make their move forward on the second night and meet the main force. Preparations then began for the move to the objective that night, but during the afternoon, the exercise director ordered the force to withdraw, as the enemy had been able to perfect his defensive arrangements, and, in accordance with the original exercise directive, the force had to be able to take part in the (imaginary) attack of the main body of the army. By this time, the infantry commander was able to use his few motorized elements to get behind the mobile force and, with the aid of friendly umpires, block the retreat with mines and antitank guns. Under these cirucmstances, the withdrawal, which was carried out in the dark, was naturally chaotic, but before Lindsay was given a chance to retrieve the situation, the exercise was terminated almost twenty-four hours before schedule.

For the opponents of the strategy of armored war, this was as great a victory as could be imagined; it seemed to show that Germains's predictions regarding the results of such an operation were correct. The successes of the earlier exercises and the restrictions placed on Lindsay were largely ignored and Liddell Hart reported that "even before the final scene, the voice of constant doubters was still to be heard, capitalizing with ill-concealed glee on the teething trouble of mechanized mobility."[122] Some protank people, apparently unaware of the situation, turned on Lindsay, placing the blame for failure on him. One especially scathing article noted that Lindsay's long career had many good marks to its credit, but that he had never commanded such a unit before and concluded, "It was taking a risk putting this both new and complicated machine in the hands of a commander who, as far as it was concerned, was an amateur."[123] Such criticism was unduly harsh, considering the restrictions Lindsay faced, but few were interested in excuses. The fact was that a mobile force operating well behind enemy lines had been shown to be vulnerable to a conventional force, and the advocates of the theory of armored warfare had suffered a severe blow.

Notes

1. J.F.C. Fuller to B. H. Liddell Hart, 7 January 1927, Liddell Hart Papers, States House, Medmenham, Marlow, Buckinghamshire, England.
2. WO File 32/2820, "Proposals for Initiation of the Experimental Mechanical Force," Minute 10, dated 11 June 1926.
3. Liddell Hart reported the subordination of the force to the Third Division and Southern Command in the *Daily Telegraph* on 14 December 1926 and the dual-command arrangements on 10 January 1927. In neither article did he voice any objections to these arrangements.
4. A copy of this letter is in Box 2, Fuller Papers, Rutgers University, New Brunswick, New Jersey.
5. Liddell Hart, on the other hand, feels Fuller was right. Sir Basil Liddell Hart, *The Tanks,* 2 vols. (New York: Frederick A. Praeger, 1959), 1:244–46 and Sir Basil Liddell Hart, *Memoirs,* 2 vols. (London: Cassell, 1965), 1: 112–18. Trythall, *"Boney" Fuller* (New Brunswick: Rutgers University Press, 1965), pp. 134–144, believes the matter was "a storm in a teacup" and that Fuller could have easily corrected it once he had assumed command of the mechanized force.
6. WO 32/2847, "Introduction of the Mobile Division." The tests involved the Third King's Own Hussars and the Ninth Queen's Royal Lancers.
7. Lieutenant General Sir Charles Broad, personal interview with the author in Beaminster, Dorset, 19–20 August 1970.
8. Liddell Hart, *Memoirs,* 1:113; Trythall, *"Boney" Fuller* p. 142, is probably correct when he interprets this as a sign Milne thought the matter was insignificant.
9. Burnett-Stuart to Fuller, 18 February 1927, Box 2, Fuller Papers, Rutger. N.J.
10. Attached to the letter is a note in Fuller's handwriting that says, "sarcastic letter from Burnett-Stuart." Ibid.
11. Quoted in Liddell Hart, *The Tanks,* 1:245 and Trythall, *"Boney Fuller,* 138.
12. J. F. C. Fuller, *Memoirs of an Unconventional Soldier,* (London: Ivor, Nicholson and Warson, Ltd. 1936), p. 439.
13. Burnett-Stuart to Fuller, 8 March 1927, Box 2, Fuller Papers, Rutgers. N.J.
14. These letters are in Box 2, Fuller Papers, Rutgers.
15. *Daily Telegraph,* 22 April 1927.
16. Liddell Hart, *Memoirs,* 1:116.
17. *Daily Telegraph,* 12 May 1927.
18. Interview with General Broad. There are three letters from Milne to Fuller in Box 2, Fuller Papers, Rutgers, dated 10 June 1927, 6 September 1927 and 6 July 1928. All three are cordial and, although the first contains criticism of how a mechanized force was handled on a recent staff side, there is no hint Milne might consider employing Fuller in this area.
19. Montgomery-Massingberd to Liddell Hart, 22 May 1927, Liddell Hart Papers. Montgomery-Massingberd is blunt in his criticism but friendly. The impression of the letter is that of a father trying to correct a son who has made a mistake.

20. "Order of 3rd Division for Collective Training Season, 1927," an undated army training bulletin, Liddell Hart papers, File A4/20.

21. Liddell Hart, *The Tanks,* 1:248-49.

22. This account is based on articles in the *Daily Telegraph,* 20 August–31 August 1927, and "Exercise Schemes for the E. M. F.," an undated, untitled series of mimeographed instructions apparently prepared by the staff of the mechanized force, Liddell Hart Papers.

23. *Daily Telegraph,* 20 August 1927. It should be noted that Liddell Hart held the title and saw himself as a "military correspondent" rather than simply a "military reporter." In this capacity, he felt his primary duty was not simply to report what had occurred, but to analyze and evaluate it. Liddell Hart, *Memoirs,* 1:80.

24. *Daily Telegraph,* 31 August 1927.

25. The summary was published in the *Daily Telegraph,* 10 September 1927. All references here are from the text in Liddell Hart Papers, File A4/20.

26. Fuller to Liddell Hart, 21 December 1927, Liddell Hart Papers.

27. *Daily Telegraph,* 16–17 September 1927; and "Exercise Schemes for the E. M. F." (see note 22 above).

28. *Daily Telegraph,* 24 and 27 September 1927.

29. R. J. Collins, "Experimental Mechanized Force," *Journal of the Royal Artillery,* 55 (April 1928): 12–26. The lecture was published by the Royal Artillery Institute.

30. The lecture was given at the University of London on 8 March 1928 and printed in the *Army Quarterly* shortly thereafter. Major General Sir John Burnett-Stuart, "The Progress of Mechanization," *Army Quarterly* 16 (April 1928): 30–51.

31. Anonymous, "Editorial," *Army Quarterly* 15 (October 1927): 1–2; 15 (January 1928): 229–30.

32. "Memorandum on Army Training: Collective Training Period, 1927," Liddell Hart Papers.

33. *Daily Telegraph,* 20 September 1927.

34. "Memorandum on Army Training: Collective Training Period, 1927," para. 5A, Liddell Hart Papers.

35. Ibid., para. 6.

36. *Daily Telegraph,* 6 and 13 July 1928.

37. This account is based on "Armoured Force Training, 1928," an unpublished training schedule for the Third Division and Armoured Force, and articles in the *Daily Telegraph.*

38. *Daily Telegraph,* 25 August 1928.

39. Ibid.

40. *Daily Telegraph,* 25 and 26 September 1928; interview with General Broad.

41. They are found in WO 32/2828, "Experimental Armoured Force Report," 1928.

42. Ibid.; Burnett-Stuart's comments are in Minute 1C and Montgomery-Massingberd's in Minute 5A.

43. Ibid., Minute 1C. He apparently felt quite strongly about this point, referring to it twice within three pages.

44. *Daily Telegraph,* 8 May 1928.

45. WO 32/2828, Minute 1C.

46. Ibid., Minute 5A.

47. Ibid., these are undated and unnumbered minutes.

48. WO 32/2825, "Experimental Formations for 1929."

49. General Broad admitted this was an error, but felt the tank advocates had no choice under the circumstance. Interview with General Broad.

50. *Daily Telegraph,* 26 February 1933.

51. Broad to Rickard, 23 June, 1970, Broad Papers, Center for Military Archives, Kings College, University of London, London, England.

52. Interview with General Pile.

53. Robin Higham, *Armed Forces in Peacetime* (Hamden, Conn.: Anchor Press, 1962), pp. 326–27.

54. *Daily Telegraph,* 16 November 1929, 21 January 1930; Pile to Liddell Hart, 16 September 1931, Liddell Hart Papers.

55. *Daily Telegraph,* 12 March 1929.

56. *Daily Telegraph,* 13 March 1930.

57. Quoted in Correlli Barnett, *The Collapse of British Power* (New York: William Morrow and Co., pp. 242–43.

58. Charles L. Mowat, *Britain Between the Wars, 1918–1940.* (Chicago: University of Chicago Press, 1963), chapter 8, pp. 413–79. A. J. P. Taylor, *English History, 1914–1945* (New York: Oxford University Press, 1965), chapter 9, pp. 298–320.

59. Robert Graves and Alan Hodge, *The Long Week-end. A Social History of Great Britain, 1918–1939* (New York: W. W. Norton, 1963), p. 304.

60. Mowat, *Britain Between the Wars,* p. 413.

61. *Daily Telegraph,* 12 April 1928. The regiment held its final horse parade on 11 April 1928.

62. Anonymous, "Editorial," *Army Quarterly* 18 (April 1929): 19.

63. All references from Great Britain, War Office, *Mechanized and Armoured Formations* (London: HMSO, 1929).

64. A copy of the speech is in file A4/20, Liddell Hart Papers.

65. WO 32/2825, "Experimental Formations for 1929," Minute 1A. The separation of light and medium tanks into distinct battalions as recommended by Broad and Milne proved to be too inflexible and was dropped in favor of mixed battalions. Eventually, the medium and light tanks were combined to form mixed companies to provide the closest cooperation between the two.

66. Ibid., Minute 71.

67. Anonymous, "Editorial," *Army Quarterly* 19 (January 1930): 232–34.

68. *Daily Telegraph,* 26 and 30 September 1929.

69. Anonymous, "Editorial," *Army Quarterly* 17 (July 1929): 226.

70. Great Britain, War Office, *Field Service Regulations,* vol. 2: *Operations* (London: HMSO, 1920), p. 22. Great Britain, War Office, *Field Service Regulations,* vol. 2: *Operations* (London: HMSO, 1924), p. 12; hereafter referred to *FSR* (1920) and *FSR* (1924).

71.Great Britain, War Office, *Field Service Regulations,* vol. 2: *Operations* (London: HMSO, 1929), pp. 10–11; hereafter referred to as *FSR* (1929).

72. *FSR* (1929), pp. 11–12.

73. Ibid., pp. 13, 17, 126.

74. Ibid., p. 13.

75. See above, p. 114.

76. *FSR* (1929), pp. 3–4, 35. The first sentence in the previous quotation from the 1920 *FSR* was retained, but the second sentence was dropped.

77. See above, pp. 114–15.

78. The exact tables of organization and equipment for these forces are found in WO 33/1033, "Composition of the B. E. F. for Small Wars, 1923"; WO 33/1058, "Composition of the B. E. F., 1924"; and WO 33/1187, "Composition of the B. E. F., 1930."

79. *Daily Telegraph,* 29 May 1930.

80. *Daily Telegraph,* 9, 25, and 26 September 1930. The quotation is from the last article.

81. WO 33/1283, "Mechanical Warfare Board: Third Report, 1930–1931," p. 8.

82. Interview with General Broad.

83. Ibid. This officer was no relation to J. F. C. Fuller.

84. Broad to Rickard, 23 June 1970, Broad Papers.

85. *Daily Telegraph,* 5 and 10 September 1931.

86. *Daily Telegraph,* 17 September 1931.

87. Broad to Rickard, 23 June 1970. He served at various posts until he retired in 1942 at age sixty. In his history of the Royal Tank Corps, Liddell Hart has a photograph of Broad briefing a group of War Office officials who, from the expression on their faces, were clearly skeptical of what he was telling them. Liddell Hart, "The Tanks, 1: facing p. 317.

88. Major General Sir Charles W. Gwynn, "Army Training, 1931," *Journal of the Royal United Service Institution,* 76 (November 1931): 722.

89. Ibid., pp. 727–29.

90. *Daily Telegraph,* 23 August 1932; Liddell Hart, *The Tanks,* 1:299.

91. The exercises were covered in the *Daily Telegraph,* 23 and 25 August, 1, 2, 3, 10, 13, 14, and 15 September 1931.

92. The committee consisted of seven officers, of whom five were general officers and two, colonels (with the temporary rank of brigadier), WO 33/1297, "Report of the Committee on the Lessons of the Great War."

93. Ibid., p. 21.

94. Ibid., p. 10.

95. Ibid., p. 11.

96. Ibid., p. 15.
97. Ibid., p. 17.
98. Ibid.
99. Ibid., p. 33.
100. *Daily Telegraph,* 20 February 1933.
101. Anonymous, "Editiorial," *Army Quarterly* 25 (January 1933): 201.
102. Field Marshal Air Archibald Montgomery-Massingberd, *Autobiography of a Gunner,* a privately published memoir in the possession of the author. My thanks go to Mr. John Montgomery-Massingberd, nephew of the late field marshal, who sent me a copy of the memoir for my use.
103. Liddell Hart, *The Tanks,* 1:301–2; *Memoirs,* 1:300.
104. P. Hobart to Liddell Hart, 24 July 1934. Liddell Hart Papers. It should be noted that Hobart was as caustic as Fuller in his private correspondence and was scarcely one to give compliments to opponents of mechanization.
105. *Daily Telegraph,* 3 October 1933.
106. The exercises were reported in the *Daily Telegraph,* 22, 23, and 31 August, 1, 5, 13, and 22 September 1933. See also "Army Training Memorandum Number 10: The Collective Training Period 1933," Addendum dated 25 January 1934, Liddell Hart Papers.
107. *Daily Telegraph,* 11 November 1933.
108. Liddell Hart, *The Tanks,* 1:306.
109. "Army Training Memorandum Number 10: The Collective Training Period, 1933," Addendum dated 25 January 1934, Liddell Hart Papers.
110. "Report on the Royal Tank Corps Staff Exercise," an unpublished memorandum, Liddell Hart Papers, A4/20, pp. 2–5.
111. Ibid.
112. Hobart to Liddell Hart, 24 July 1934, Liddell Hart Papers.
113. Liddell Hart, *The Tanks,* 1:317. The variety was due to continued uncertainty as to how the light-tank battalion should be organized.
114. Hobart to Liddell Hart, July 24, 1934, Liddell Hart Papers. Hobart is quoting from "Army Training Memorandum Number 10: The Collective Training Period, 1933," para. 3(1). The original could not be found.
115. Wavell to Liddell Hart, 27 May 1948, Liddell Hart Papers.
116. *Daily Telegraph,* 28 August 1934.
117. Hobart to Liddell Hart, 7 October 1934, Liddell Hart Papers. Hobart wrote, "Of course . . . it was a complete "frame up" to raise the morale of the infantry and cavalry," but also expressed the opinion that this was necessary and, under the circumstances, good. Wavell, who was the chief umpire during the exercises, disagreed that the exercise scheme was weighted unfairly against the mobile force, but did admit that Burnett-Stuart wanted "to correct some of the wild talk that was going on about the infantry being out of date." Wavell to Liddell Hart, 27 May 1948, Liddell Hart Papers.
118. This account of the exercises is based on Liddell Hart, *The Tanks,* 1:328–35; *Daily Telegraph,* 24 September 1934; A. G. Cunningham, "The Training of the Army, 1934," *Journal of the Royal United Service Institution* 79 (November 1934): 723–32; and I. R. Kennedy, "Salisbury Plain Maneuvers," *Army, Navy and Air Force Gazette* 75 (pt. 1, 27 September 1934; pt. 2, 4 October 1934): 767–70, 787–89.
119. Kennedy, "*Salisbury Plain Maneuvers,*" p. 768.
120. Liddell Hart, *The Tanks,* 1:336.
121. Hobart to Liddell Hart, 24 October 1934, Liddell Hart Papers.
122. *Daily Telegraph,* 24 September 1934. The earlier exercises were passed off by one writer with the brief comment that "in its early training considerable success was obtained in control and flexibility" Cunningham, "The Training of the Army," p. 731.
123. Kennedy, "Salisbury Plain Maneuvers," p. 788.

• 6 •
The Period of Early Rearmament: 1934–37

Introduction

THE three years following the 1934 tank exercises may be termed the "renaissance" of the British armed forces of the interwar years. On March 4, 1935, the government issued a white paper on defense in which it announced that the deteriorating international situation required that the nation undertake a limited rearmament program. For the first time since the end of World War I, the armed forces were given sufficient funds to begin correcting the many deficiencies that had been allowed to develop. The army benefited from this program, but to a lesser extent than the other services. The requirements for aerial and naval rearmament were obvious in light of German rearmament, but the need for army expansion was not. With few exceptions, government leaders refused to accept the argument that the Low Countries were vital to the defense of Great Britain and that British ground troops were essential to the defense of the area.[1] As a result, while the army began to plan for expansion and possible employment on the Continent, its material condition continued to go wanting.

This state of affairs was paralleled within the armored forces. During this period the army decided to replace the cavalry division of the expeditionary force with a "Mobile Division" and to mechanize the remaining cavalry regiments of the regular army. Consequently, they doubled the size of the Royal Tank Corps to provide the necessary units for this new division as well as the tank battalions for infantry support and agreed to a construction program of some 2,000 tanks. Yet, these changes remained largely on paper, so that by 1937 the armored forces were only slightly larger and better equipped than they had been three years earlier. More important, the army's ideas of how to employ armored forces remained static. The cavalry concept, which, in 1934, was a progressive step over the doctrines of the previous decade, was, by 1937, as obsolete as the old Vickers mediums, which still formed the backbone of the Tank Brigade.

The Decision to Mechanize the Cavalry

Despite the shadow cast by the last exercise of the 1934 training season, the army moved quickly to replace the Cavalry Division with a wholly mechanized unit. On October 15, 1934, Montgomery-Massingberd raised the issue in a memorandum to the other members of the Army Council in which he declared:

> I have reached the provisional conclusion that, instead of having an independent Tank Brigade and a Cavalry Division containing two horsed Cavalry Brigades, a more suitable organization would be a "Mobile Division" consisting of the Tank Brigade and one mechanized cavalry brigade.[2]

That Montgomery-Massingberd should issue such a proposal is less surprising than might appear at first glance. Although he had been and would remain a staunch opponent of the strategy of armored war, he had shown an awareness of the advantages of mechanizing the cavalry at least since the report of the Cavalry Committee he had chaired eight years before. Moreover, the problem of obtaining horses to support the army had, by this time, reached critical proportions. Indeed, in early 1935 a writer in the *Journal of the Royal United Service Institution* estimated that, to support an expeditionary force of twelve divisions, Great Britain would have to import 6,500 horses a month for the first six months and 8,000 horses a month thereafter.[3]

Montgomery-Massingberd's proposal immediately raised the question of who would control the expanding tank forces. With the demise of the horse becoming increasingly apparent, many cavalry officers now urged their units be reequipped with light tanks or armored cars lest their branch be faced with virtual extinction. They were able to muster strong support for such a course. The preservation of regimental spirit, the esprit de corps of the cavalry and the social prestige of cavalry officers were all considerations sure to appeal to large segments of the army and were widely used.[4] They also advanced arguments, however, that were based on non-traditional grounds. The most important of these was that the replacement of cavalry horses with light tanks and armored cars was simply a change from one mode of transportation to another and that the units concerned would carry out the same missions. Both Fuller and Liddell Hart, in fact, had used this argument in the 1920s to win adherents among the cavalry. Fuller had stressed this in his articles on the future of cavalry in the *Cavalry Journal* in 1920,[5] and Liddell Hart repeated this theme on numerous occasions. When the Eleventh Hussars was mechanized in 1928, for example, he lamented the cavalry's decline on the modern battlefield and called on it to recognize that only through change such as this could its former position of prominence be restored.[6] Many also argued that cavalry officers, by virtue of their training, were ideally suited to command tank units. The key to success in the operations of both types of units was mobility and cavalry officers were depicted as incipient tank officers because of their training in this form of warfare. This view also had found favor with Liddell Hart in the late 1920s, who had written at that time:

> It is the cavalry man's training which has given him the instinct of maneuver and rapid decision, and the increasing vulnerability of his mount has helped to develop these instincts. The cavalryman, too, has escaped most of the "arthritic" effect of trench warfare. In the last war, tactics submerged strategy, and half of the troubles of today are due to this lost sense of strategy. Only with the cavalryman does his training keep it alive and prevent an excess of "blood" in the head. The best hope of using armoured vehicles aright is to put more cavalrymen in charge of them. The infantryman or gunner, gifted with unusual imagination, may acquire the art. The good cavalryman is, by his past training, a ready-made artist.[7]

Clearly, if this were the case, there was considerable logic to mechanizing the cavalry rather than expanding the Royal Tank Corps.

Not all tank advocates agreed with these positions, however. Liddell Hart's article had provoked a strong letter of protest from Charles Broad, who argued that the cavalry had no idea of how to work with mechanical vehicles, that it was mobility plus firepower that gave tanks their great superiority, and that since the cavalry had no experience with the latter, their value in tank units would be severely limited.[8] By the mid-1930s the tank advocates' position had solidified: they did not object to giving armored cars to the cavalry, but they were adamant that the Royal Tank Corps should have control over all tanks. Their position was most completely presented in a paper entitled "The Organization of Tactical Vehicles" written by Hobart and Lindsay in early 1935. The essential consideration in deciding the division between the cavalry and the Royal Tank Corps, they argued, was not the role of either arm, but, rather, the need to retain uniformity in the equipment each used. Tanks differed from other types of vehicles in that they were expected to "enter the hostile position" during combat, and this role required they have certain characteristics. Specifically, they should be fully armored, have all-around-traverse of weapons, and be able to fire on the move, move rapidly across country, and be controlled by radio. A vehicle embodying all of these characteristics would be both expensive and extremely difficult to operate and, they concluded, should be under the direction of a single control, meaning, of course, the Royal Tank Corps. Lindsay and Hobart compared the situation facing the tanks to that facing the heavy machine gun during World War I. They noted that administrative and training needs rather than tactical requirements had led to the formation of a separate Machine Gun Corps, and they argued that the same reasoning should apply in this instance. To put it simply, the tank required a high degree of technical proficiency, which only the Royal Tank Corps possessed, and to place it in the hands of another branch of service that lacked the corps's expertise would only complicate training and increase the chances that the weapon would be misused. The existing units of the Royal Tank Corps would have to supply the men to train the new tank units no matter how they were formed, and it would only be logical to keep the entire process under one roof.[9]

These considerations were reinforced by the growing feeling among the tank advocates that, since they had done all of the groundwork, they should be allowed to reap the fruits of their endeavors and, equally important, that they were the only ones who knew how to handle tanks correctly. By 1935, Liddell Hart had grown skeptical of the cavalry's ability to handle tanks and, reversing his earlier position, wrote:

> But some doubt must remain, until experience has proven the contrary, as to the ability of horse-minded soldiers to become mechanically minded, and to get as much out of their machines as a corps of natural enthusiasts for machines. Past experience, as a whole, does not lessen this doubt.[10]

The War Office decision to mechanize the cavalry seems to have been a foregone conclusion; indeed, the available records contain virtually nothing concerning the actual decision to equip the cavalry with light tanks or an explanation of why it was made. From the papers available, however, two threads emerge that give some indication of the army's thinking. The first was its belief that mechanized cavalry and the Royal Tank Corps would fulfill different roles. In its view, the mechanized cavalry "will be carrying out their traditional functions of reconnaissance and protection while that of the tank battalion will be mainly offensive."[11] To put it in traditional terms, the Royal Tank Corps would replace the old heavy cavalry while the mechanized cavalry would replace the old light cavalry. Given this assumption, the equipment and organization of the two arms would also allow for clear division. The Royal Tank Corps would be equipped primarily with medium and heavy tanks organized into army tank brigades and infantry-support battalions. They would add to the offensive power of the army either by independent flanking attacks or by direct support of the infantry. The mechanized cavalry, on the other hand, would be equipped with light tanks and armored cars and conduct reconnaissance and screening actions. This division of function would not solve all of the administrative and training problems that Hobart and Lindsay had envisaged, but it was, nonetheless, a system possessed of a considerable degree of logic.

The second reason for the army's decision was undoubtedly the pressure of social considerations. As the cavalry formed the social elite of the British Army (although not, as we have seen, the professional elite), it was natural for the army to desire to preserve its distinctiveness. The extent to which it would go to achieve this end was most clearly demonstrated in 1939, when the army was considering ways to amalgamate the Royal Tank Corps and the mechanized cavalry regiments into a single body to be known as the Royal Armoured Corps. The result was a decision made in a way that would preserve the position of the cavalry regiments: the new corps would be composed of the Royal Tank Corps, henceforth designated the Royal Tank Regiment, and the eighteen mechanized cavalry regiments in which the enlisted personnel, but not the officers, of the new corps would be interchangeable. Both the Royal Tank Corps and each of the cavalry regiments would maintain their own list of officers. The deciding argument in this decision was that, since cavalry officers, unlike tank officers, generally had a private income, and since the cavalry regiments were generally more expensive, it would be unfair to all officers concerned to mix the two. Either the cavalry officers would have to reduce their standard of living or the tank officers would have to stretch their means to the breaking point.[12] In view of the fact that the committee reported this was a widespread feeling—especially among the cavalry—it can be assumed that it was also a consideration in 1934 and 1935 when the question arose as to how to organize the army's armored forces.

Both at this time and in later years, Liddell Hart decried the decision to mechanize the cavalry as a victory for the less progressive forces in the army. The War Office, finally backed into a corner by the dwindling supply of horses, he argued, had been forced to accept the mechanization of the cavalry; but it still refused to put the new arm completely in the hands of the Royal Tank Corps, as it feared that "the pace of progress might become too fast to be controllable."[13] That there were those in the War Office who harbored such fears was undoubtedly true, and the effort to preserve the social elitism of the cavalry demonstrated this. But it is no less true that the organization of the armored forces established in 1935 was a logical extension of the "cavalry concept" of tank warfare, which the army had already accepted. Moreover, Liddell Hart himself had strongly advocated the mechanization of the cavalry just a few years previously and had exalted the cavalry officer as the ideal tank officer. It would be less than charitable to ascribe entirely different motives to the senior War Office officials in 1934. The crux of the matter was that the army's strategic doctrines and its social values went hand in hand in this instance. That the army was happy this was the case is undoubtedly true, but the decision cannot be ascribed simply to the desire to protect the cavalry.

The consequences of this decision, as it affected the development of the armored forces, were twofold. In the first place, it retarded the development of the necessary technical proficiency in the newly mechanized units. This was such a problem, in fact, that a committee studying the Royal Armoured Corps in 1939 noted that the cavalry still needed the assistance of Royal Tank Regiment personnel and hoped that steps would be taken to better facilitate the training of the mechanized cavalry.[14] The second result was that it placed former Tank Corps officers in competition with cavalry officers for senior positions at a time when this was distinctly disadvantageous to the former. The Royal Tank Corps was a relatively young service and, as a result, had very few senior officers. The only two exceptions were J. F. C. Fuller, who had retired in 1933, and George Lindsay, whose career was still under the cloud of the 1934 exercises. As for the remainder, it would be several years before they could exert any influence over army policy.

Course of Development, 1934–37

The decision to initiate trials to determine the organization of the mechanized cavalry regiments of the Mobile Division was made public on November 14, 1934, as part of an overall plan of army modernization. In addition to the formation of the Mobile Division, plans were announced for the mechanization of all infantry division vehicles, the addition of support weapons in these divisions, the mechanization of the light artillery brigades, and the expansion of the Royal Tank Corps to form one army tank battalion for each of the infantry divisions in addition to the mixed and light-tank battalions for the Tank Brigade.[15] Significantly the impetus for these changes was the growing possibility of a Continental commitment. The Ten-Year Rule had been rescinded in 1932, and, following the failure of the disarmament conferences in 1933, the War Office created the Field Force Committee to study its requirements for a Continental war. In June 1934, the committee recommended that the Regular Army should be capable of being dispatched within one month of the outbreak of hostilities, with the Territorial Divisions to follow as soon as possible. This was obviously impossible given the existing state of the army, however, and Montgomery-Massingberd specifically linked the modernization program to the army's new role in his October memorandum when he stated that the change was being made "in light of the Western Plan."[16] Yet this growing possibility that the army would have to fight on the Continent did not change its strategic doctrines or the role of its armored forces within them. The 1934 program, with its emphasis on the mechanization of the infantry and artillery was clearly designed to modernize the army along traditional lines rather than radically alter its essential character, that is, to enable it to conduct a strategy of attrition more efficiently rather than to implement the theory of armored warfare.

This was even more clearly reflected in a paper written by Montgomery-

Massingberd in September 1935. Entitled "Future Organization of the British Army," its purpose was "to review, on broad lines, the organization of the British Army with particular reference to the modernization of the Field Force."[17] The paper was noteworthy not only for the clarity of its presentation, but also for the progressiveness of the views it espoused, for despite Montgomery-Massingberd's refusal to accept Fuller and Liddell Hart's ideas on how armored forces should be employed he was convinced that such forces would play a crucial role in a highly mechanized and mobile expeditionary force. His general theme was that any future European war would follow the same strategic lines as the Great War. It would open with a repetition of the German invasion of France, this time through Holland as well as Belgium to increase the chances of outflanking the Allies. He stressed that, under these circumstances, Great Britain's first line of defense would have to be the Low Countries, for they not only would provide the bases for Germany to launch air attacks against England, but also would provide England with the bases to attack Germany. Regarding the actual conduct of operations, he predicted that the war would pass through three stages: a highly mobile initial stage, a larger period of static war, and, finally, a period of breakthrough and exploitation by one side. Mobility would obviously be the key during the first stage, and, in a passage almost reminiscent of Fuller and Liddell Hart, he wrote:

> A small force which is better trained, better organized, better equipped and more mobile than the armies of the Continental Powers is likely to have a greater moral and material influence than a much larger force which is less efficient.[18]

This passage clearly sums up the intent and goals of the reforms he was proposing at this time. They were not designed to increase the size of the army—he noted the need to reduce the size of infantry divisions to make them more flexible and mobile—but to increase its mobility and firepower. Within this overall framework, the duties of the Mobile Division were to be basically that of the cavalry—primarily reconnaissance and screening actions—but he added that it would also be of great value in flanking maneuvers in conjunction with an assault by the main body as well as in exploitation after "the enemy's main resistance has been broken down."[9]

Assuming this mobile phase did not prove decisive, Montgomery-Massingberd predicted the war would then enter a static phase similar to the period of trench warfare of World War I. During this period, which he believed would last from eight to twelve months, the size of the British Army would be significantly increased by raising Territorial Divisions and going into full production of the latest types of armored fighting vehicles. Meanwhile, the army would remain strategically on the defensive, but could engage in limited tactical offensives.[20] When the program of expansion was completed, the army would then go on the strategic offensive.

Significantly, he realized the great artillery bombardments characteristic of the Western Front had done little to breach the German defenses. In the future, he believed they should be replaced by tanks, which could not only guide the infantry through the enemy's defenses, but also would maintain the momentum of the exploitation once the breakthrough had been achieved. Indeed, noting the failure to maintain this momentum on the Western Front, he wrote that "it seems reasonable to assume that suitable mechanized and armoured formations, acting in combination with aircraft, could have achieved it."[21]

As a whole, Montgomery-Massingberd's views were a curious blend of both progressive and traditional ideas. His view of the future of the Regular Army was extremely progressive, but curiously, he did not extend it to the expanded army. He also recognized the futility of the offensive methods of the Western Front and the value of tanks not only as infantry support weapons, but also on independent operations, but he failed to see the decisive impact tanks could have if they were made the center of gravity of the army as a whole. In a particularly curious reference, he also noted that the German successes on the Western Front in 1918 were due largely to new infantry tactics that had "perhaps not been fully recognized," but he provided no details of these tactics, which, it will be recalled, had formed the basis for Fuller's Plan 1919.[22] Despite these shortcomings, however, Montgomery-Massingberd's plan would have established the British Army as the most highly mechanized army in the world and the one with the highest proportion of tanks.

If Montgomery-Massingberd's plans were generally progressive, however, his means were determinedly methodical and cautious. In part, this was a product of the domestic and foreign situation facing the army in 1934: there was, after all, no threat from Nazi Germany at this time, and, as Montgomery-Massingberd was aware, British public opinion would not have tolerated any program calling for rapid rearmament. But it was also a product of his own character and his long experience as a staff officer, which bred a high degree of caution. As he wrote:

> The more one goes into that problem—mechanization—the more one realizes that, unless we go cautiously, we shall find ourselves in difficulties. First of all, if we mechanize too much, we may, as he [MacReady] told you, build up an enormous tail; and, secondly, if we mechanize too quickly, we may find that in a year or two the larger part of the tanks or machine gun carriers, or whatever the vehicles are, may very probably be out of date. I am therefore quite convinced in my own mind that what we have to do is go along, going into production to a limited extent whenever we have found a really good article and whenever we can get the money, and at the same time continue to experiment as hard as we can with as simple forms of vehicles as we can find so that if war looms on the horizon, we may be in a position to go into

production at once with the latest pattern on the stocks. I am quite certain this is the only way we can carry on with our limited resources.[23]

This should not be construed to mean that he was halfhearted concerning his program, however. The War Office papers relating to the replacement of the cavalry division with the Mobile Division in the Expeditionary Force clearly demonstrate that it was Montgomery-Massingberd who was the moving force behind this decision. The other members of the Army Council, while not overtly objecting to the proposal, were far more concerned with the difficulties it presented. The AG was particularly pessimistic, stressing, among other considerations, the problem of balancing mechanized and horsed cavalry regiments at home and abroad. To this, Montgomery-Massingberd curtly replied that there would be a sufficient number of horsed regiments to maintain the system for the present; as for the future, he wanted to mechanize more units abroad so this would also not be a problem.[24] Thus, if his pace was slow and methodical, it was also forceful and determined.

The last months of 1934 were devoted largely to resolving the details of the organization and role of the Mobile Division. As tentatively planned, the combat portions of the division would consist of a reconnaissance group of two armored-car regiments, the Tank Brigade of one light-tank and three mixed-tank battalions, and a mechanized-cavalry brigade composed of an armored-cavalry regiment equipped with light tanks and three motorized-cavalry regiments.[25] The division was without any infantry support, for, although the motorized-cavalry units were, in fact, motorized infantry, their duty was close reconnaissance. This absence of infantry was another indication that the cavalry concept was dominating the development of the armored forces, and this was confirmed when the duties of the Mobile Division were specified in an Army Training Memorandum of December 1934. According to the memorandum, these duties were (1) to reconnoiter and protect the main body of the army during movement; (2) to delay the enemy, safeguard a flank, form a mobile reserve or cover a withdrawal during battle; (3) to exploit a success and pursue the enemy following a victory; (4) to support the main attack on or close to the battlefield (to be done mainly by the tank brigade); and (5) to conduct special missions such as raids.[26] No clearer statement of the cavalry concept could have been formulated.

The spring of 1935 was a period of considerable activity for the army. Plans were made to hold maneuvers from late August to mid-September, the first since 1925, and tests were to be conducted to determine the exact organization of the motorized-cavalry regiments of the Mobile Division. The Third King's Own Hussars had volunteered to undergo temporary mechanization for the tests and were being equipped with the necessary vehicles to conduct the experiments.[27] Much of the impetus for these activities was provided by the knowledge that the army was getting additional funds. The army estimates for 1935 provided for an increase of £ four million, but the financial stringency of the previous few years had been so severe that most of the new funds had to be allocated to replace depleted ammunition stores and new equipment for the infantry and artillery. Consequently, only an additional £500,000 was actually available for mechanization throughout the army, bringing the total to £1.368 million. This small increase was hardly satisfactory to the tank advocates, but in view of the situation in which the army found itself, no alternative was possible.[28]

Meanwhile, the First Tank Brigade was laying plans for its second year of training. During the previous winter, one of the brigade's greatest problems had been largely eliminated when the number of its support vehicles was reduced from 217 to only 99. As Liddell Hart noted at the time, such reductions were vital if mobile forces were to achieve their full potential and also illustrated that mechanization need not necessarily produce an impossibly large logistical element.[29] Less heartening to the tank advocates was the news that, while the Tank Brigade would conduct several brigade-scale exercises in August, it would be broken up for the final exercises in September to act as army-tank battalions. Naturally, they believed the brigade should remain together as this was its primary role and viewed the decision as a step backward. In reality, however, it was simply another consequence of the army's lack of tanks. The Army Training Memorandum of the previous December had clearly emphasized the independent role of the tank brigade and its value on the outcome of operations. It had also described the need for army-tank battalions to support the infantry, however, and the need for training in this type of operation as well as the former. Since there were no army-tank battalions, the brigade would have to fulfill both roles until the new units could be raised and this was the reason for the decision to break up the Tank Brigade for the final exercise.[30] It was an unfortunate situation, but from the army's point of view it clearly was not designed as a step backward.

The 1935 army maneuvers began on August 22 with a series of brief exercises by the Tank Brigade, the first being a recreation of the battle of the Marne. The brigade was assigned the mission of penetrating the breach between advancing enemy armies and disrupting their rear areas so that they could not effectively parry a blow to their front. The tanks conducted a fifty-mile approach march with ease and successfully completed their mission. On August 26, another exercise on the same basic pattern was conducted. In this instance, the tanks penetrated to the enemy's rear and attacked an artillery area in conjunction with an assault by the main body of the army. Again they were successful.[31] Although both exercises involved the Tank Brigade acting as an independent unit and maneuvering over large distances, it should be stressed that both were based upon the cavalry concept of tank warfare rather than the theory of armored warfare. In neither case were the tank attacks designed to achieve a true strategic penetration behind enemy lines or to be the decisive blow.

The final exercise of the 1935 season was the least interesting despite the fact that it was the largest held in a decade. The one major change from that earlier year was the evidence of some modern equipment. The engineer, signal, and supply units all had been mechanized, as had most of the artillery, but much remained to be done. The participating forces were divided into two equal groups consisting of two infantry divisions, a cavalry brigade, two tank battalions, and supporting troops. The exercise itself revealed nothing either old or new. The two forces advanced on each other in parallel lines and Deverell, commander of the "Eastern Force," was able to pin down one of his opponents' flanks and then turn the other. The maneuver was the only one in which tanks did not play a major role, as they were tied down to infantry support.[32]

Far less publicized were the mechanized cavalry exercises involving the Third King's Own Hussars. This regiment, it will be remembered, had been converted into what was in reality a motorized-infantry unit but was given the mission of determining the extent to which it could replace horsed cavalry.[33] These orders were taken quite literally by the commander of the Third Hussars, Lieutenant Colonel Grubb, who later admitted:

> Every effort had been made throughout to keep the tactics of mechanized cavalry to the principles laid down in Cavalry Training, vol. II with the motor vehicle substituted for the horse. The idea has been to evolve mechanized cavalry out of horsed cavalry as opposed to doing something new. All the principles of cavalry action have been simulated, and the tactical operations of horsed cavalry have been carried out only with vehicles substituted for horses.[34]

Nonetheless, Grubb concluded following the summer's training that this new form of cavalry undoubtedly could replace the old. Although he felt that close cooperation with armored cars and light tanks was necessary, he asserted in the strongest terms that the tests had demonstrated the ability of motorized cavalry to accomplish all that horsed cavalry could with far greater ease and efficiency. In his own words:

> Mechanized cavalry by reasons of their speed and ability to travel far without fatigue are more efficient than horsed cavalry, when sent ahead to seize positions. They can also hold positions more effectively owing to their superior firepower. Mechanized cavalry can picket a flank or the rear more efficiently than horsed cavalry for similar reasons. Mechanized cavalry will be far superior to horsed cavalry in the pursuit because of their greater mobility and ability to carry machine guns. A very marked feature of mechanized cavalry is the freshness of personnel carried in the section cars compared to horsed cavalry or infantry after a long march.[35]

THE MOBILE DIVISION 1935

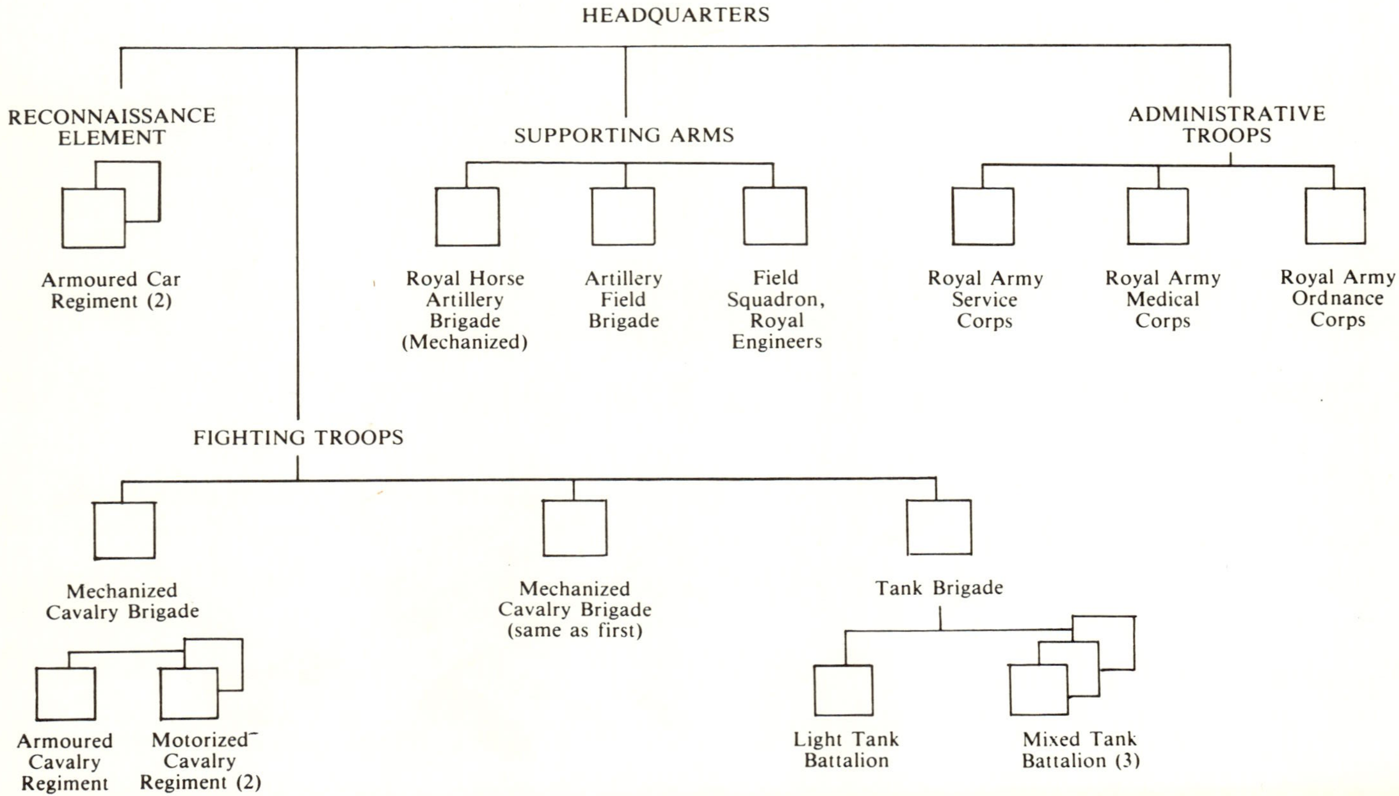

Grubb's superiors generally concurred with these views, although they differed with him on some of the details. Some felt there still might be a need for horsed cavalry in certain types of terrain, but there was no question about the necessity for mechanized cavalry in the Mobile Division or that such a force would prove to be of considerable value.[36] It is noteworthy, however, that no one questioned the validity of the motorizing cavalry in such a manner or asked how such cavalry would differ from motorized infantry.[37]

During the fall of 1935, further efforts were made to bring the Mobile Division closer to reality. Montgomery-Massingberd had visited France the previous summer, and, impressed by what he saw of a French mechanized-cavalry division on maneuvers, returned with a determination that such a division be formed in England as soon as possible. Shortly thereafter, the Army Council formally agreed to replace the cavalry division with a Mobile Division,[38] and Montgomery-Massingberd set about to win governmental approval for the change. He was particularly concerned that time should not be lost because the training of officers for the division would take some time after it was formed and the unit would obviously be one of the first to go abroad in case of war. As he wrote the secretary of state for war,

> I do feel that we should be getting on with our Mobile Division as quickly as we can, and the sooner we can appoint a commander to supervise its instruction and [the] necessary changes in organization the better. That such a formation is a powerful engine of war I have not the slightest doubt after what I have seen and it would be quite invaluable at the beginning of a campaign.[39]

Montgomery-Massingberd took more positive action on October 15, when he submitted a definite proposal for the formation of the Mobile Division to the Army Council. In contrast to his proposal of the previous year, he now suggested that the mechanized-cavalry brigade be organized into two smaller brigades rather than a single larger one. Although the change was designed to make handling the brigade easier and allow for more flexibility, it would also enable the army to maintain the balance of horsed and mechanized units at home and abroad.[40] Accordingly, he now proposed the mechanization of eight cavalry regiments, leaving the cavalry ten mechanized regiments and ten horsed regiments. The horsed regiments could not be further reduced until India agreed to do the same, and, while Montgomery-Massingberd felt this should be done, he was disinclined to push the matter. As he told the council, "I consider that there is still a need for divisional or corps cavalry regiments to work with the Field Force and we also need Army cavalry formations and units for use in terrain unsuited to mechanized troops."[41] The Army Council's reaction to this proposal was entirely positive, an indication, quite possibly, of Montgomery-Massingberd's dominance of his military colleagues rather than a change of heart on their part. The AG (who was General Sir Harry Knox, an outspoken traditionalist) expressed the opinion that the new division "will be a valuable formation"; the QMG commented he was "confident such a formation is essential for our Field Force"; and the MGO wrote, "The formation of the mobile division is a very necessary step to take."[42] On December 18, 1935, the council officially announced its decision to form the Mobile Division as soon as possible and to mechanize the eight cavalry regiments the division required according to three patterns: armored-car regiments, motorized-cavalry regiments, and armored-cavalry (light-tank) regiments.

The new status these decisions gave to the armored forces was reenforced by the 1935 revision of the *Field Service Regulations,* which, while not radically different from the previous edition, described their power in stronger terms than had hitherto been used. In its description of tank brigades, for example, it stated:

> The wide range of action and the mobility of tank brigades, *greater than any troops have ever before possessed,* give them *frequent opportunities* for surprise, and enable them to strike a blow not only at the flanks of an enemy but also at his headquarters and rear services. Even the mere threat of their attack may paralyse or hamper enemy movement by compelling him to employ an undue proportion of his force in the protection of his flanks and rear.[43]

As for the army tank battalions:

> They are *required* to break down wire entanglement and to destroy or neutralize machine guns. They are a *valuable* aid to gaining the advantages of surprise and initial success in the attack and to maintaining its momentum.[44]

Yet these changes did not signify any alteration of the army's basic strategic doctrines. The decisiveness of battle was expressed in sightly less strident terms—it was now described as "usually the decisive act in land warfare"—and it admitted, "It may occasionally happen that successful maneuver enables the objective to be attained without serious fighting," but too much cannot be read into these qualifications.[45] The tenor of the manual makes it clear that the army still viewed battles as the most important act in war, and that it believed they would conform to the general pattern that had been set forth by the Napoleonic model. Although it stated that "no fixed rules for battle can be laid down," it held that battle would consist of three phases: the preparatory stage, during which the enemy was immobilized to a fixed position; the main attack, designed to penetrate the enemy's line at a point that would make his entire position untenable; and the pursuit, during which the enemy's retreating forces would be harassed and prevented from reestablishing a new position.[46]

Despite the initial disclaimer, there is no doubt that the army perceived these stages as occurring successively, and that, in fact, its view of battle had really not changed at all from that expressed in the earlier postwar editions of the *Regulations.*Within this model, independent tank operations as described above could be of considerable assistance, but never could they be decisive by themselves. Nor could they significantly alter these stages of battle. In the words of the *Regulations:* "The employment of mobile and armoured forces in the battle will usually mark a decisive phase; the tasks allotted to them should therefore be designed to ensure, or confirm, the success of the main attack."[47]

Further development of the Mobile Division passed from Montgomery-Massingberd's hands at this point, for he retired in February 1936. His tenure as CIGS still had a year to go, but as he and Milne had held the post together for a decade, he felt the change was necessary "in order to facilitate the promotion of younger officers."[48] Before leaving, he drew up a paper for his successor entitled "Handing Over Notes," in which he outlined his personal view of the state of the army and the changes he had been able to achieve as CIGS.[49] He opened by noting that his "biggest battle" had been against the movement toward the policy of "limited liability." Its effect, he wrote, was to complicate both the training and the equipping of the army, not to mention the fact that he strongly felt the policy was a serious error. With a foresight many lacked at this time, he realized that control of the airbases in the Low Countries should be Great Britain's first line of defense, and that this could only be done by air and ground forces on the Continent.

Despite this battle, Montgomery-Massingberd believed he had achieved considerable progress in the modernization of the army. Notably, when listing his achievements, he gave first priority to those which dealt specifically with the development of Great Britain's armored forces: the establishment on paper of the Mobile Division; the establishment of the First Tank Brigade as a permanent formation; and the drastic reorganization of the cavalry calling for the mechanization of eight regiments. Of this plan, however, only the mechanization of one cavalry regiment was to be completed during that year and the last would not be equipped for another two years. He also noted that the principle of army-tank battalions had been approved. In the other arms, he reported considerable progress with the mechanization of artillery and infantry transport, and he concluded that this process would be completed in the infantry by the end of the year.

There can be little doubt that Montgomery-Massingberd felt that this was a noteworthy list of accomplishments in 1936 and, indeed, could see no reason to alter this evaluation toward the end of his life. Much was still left undone, but for that he blamed the government. Equipment could not be procured without government funds, and these simply had not been provided. He was sure in his own mind that the best possible use had been made of the funds available to the army, and that the modernization achieved was the maximum under the existing circumstances.[50] Although it is difficult to accept this evaluation in its entirety—his belief that horsed cavalry still had a limited role in war and his adament refusal to accept the theory of armored war are the most glaring contradictions—it is clear that he was by no means a rigid defender of traditional interests. The War Office papers demonstrate quite clearly that he was the moving force behind the inception of the Mobile Division and that he moved on this matter with forcefulness and determination. His pace of modernization was too slow to prepare the army for a major war in 1939, but to foresee this eventuality in 1935 would be to expect too much, and, furthermore, it would have been useless, given government policy. Whatever may be said of what he failed to do, it must be stated that he accomplished far more in this three years than had been accomplished in the previous fifteen. Given the state of Great Britain and the continued government apathy toward the army, this was no mean achievement.

Montgomery-Massingberd's successor as CIGS was General Sir Cyril Deverell. Like his predecessor, Deverell was sixty-two when he assumed this post, but he was more widely known. He had entered the army in 1895 as an infantry officer, and his career followed the traditional pattern of tours in England and abroad until the outbreak of World War I. During the war, he earned the reputation of a ruthlessly aggressive combat commander, and he rose from captain to major general in only two years. After the war, he was stationed in India, where he held various administrative posts from 1921 to 1931, and then returned to England to take over the Eastern Command. He was soon considered one of the leading contenders to succeed Montgomery-Massingberd as CIGS and his "victory" in the 1935 maneuvers was undoubtedly a major factor in his appointment.

Any assessment of Deverell's policies regarding the development of Great Britain's armored forces must take account of the growing rift between the War Office and the tank advocates over the army's rearmament policy. The War Office plan, the basis of which had been developed by Montgomery-Massingberd in late 1934, called for the mechanization of all army vehicles and the modernization of the equipment of all branches. Within this framework, the expansion and reequipment of the Royal Tank Corps received considerable emphasis, but it remained only one element in an overall plan of modernization. To the tank advocates, this was obviously a waste of time and effort: they were convinced that the only correct course of modernization lay through the expansion of the Royal Tank Corps. Liddell Hart spelled out their criticism:

> The recent progress of the Army had been mainly directed to improving the general mobility of the other arms; this is rather in contrast to the latest trend abroad, where efforts are being made to create a number of

> "mechanized" divisions in the full sense, basically composed of armoured fighting vehicles. Here the one tank brigade has continued in existence, but is not yet completely equipped with up-to-date light tanks and no medium tanks.[51]

These conflicting views were the logical outgrowth of the continuing controversy between the strategy of attrition and the theory of armored war. Quite simply, each side was looking to different elements within the army to provide the decisive punch and naturally sought to place those elements at the center of the army's rearmament program. Unfortunately, the tank advocates severely weakened their case by failing to see the need for mobile infantry to support armored formations and by their call for the creation of an elite, mechanized element within the army. Both of these positions were not only politically unwise, but also militarily unsound. They could not help but alienate many officers outside the Royal Tank Corps. Moreover, the comparison of British modernization plans with those of Continental armies simply was not applicable because of the small size of the British Regular Army. The Germans, for example, were creating an elite mechanized force within their army that, by 1940, would account for one-eighth of their total strength; but because of the total size of their army, this elite force would be more than three times the size of the entire prewar British Regular Army.[52] In short, the British Army was too small to have an elite force within itself. One is forced to return to the conclusion that, despite its adherence to the strategy of attrition, Montgomery-Massingberd's plan for the modernization of the army would have gone a long way toward laying the basis for further progress had his successors continued to move forward.

Unfortunately, Deverell was not the man to continue this progress. He has been accused, in fact, of opposing even the cavalry concept of tank warfare and of seeing the tank as nothing more than an infantry-support weapon,[53] and, while this is probably an exaggeration, the development of independent armored forces did not continue at the same pace under his direction as it had under his predecessor's. The Mobile Division, for example, was not actually formed until the end of 1937, despite Montgomery-Massingberd's continued insistence on its formation two years earlier; and then, Deverell adamantly opposed giving it to one of the tank experts, whereas Montgomery-Massingberd had wanted to give it to Lindsay. In short, while Deverell made no attempt to undo the work of his predecessor, it is clear that he did not place the same emphasis on the development of independent armored forces.[54] In light of the fact that the threat from Nazi Germany was becoming increasingly obvious during this period, this failure appears even more indefensible.

The army's attitude was reflected in its allocation of funds from the 1936 army estimates. For the first time since the end of the war, the government had granted the army a significant increase in funds so that it could begin replacing its old and worn-out equipment.[55] Most of the increase, however, went to the traditional arms for mechanization of the infantry's vehicles and large-scale mechanization of the cavalry and artillery. Little was allotted to the Royal Tank Corps, whose mainstay, the Vickers medium, was now twelve years old.[56] The situation in the First Tank Brigade was made even worse when it was stripped of its most modern piece of equipment, the Mark VI light tank, to reenforce the British forces in Egypt during the Ethiopian crisis. Hobart undoubtedly expressed the depression of many when he wrote to Liddell Hart that "bricks without straw is a normal task—but one must have some clay."[57]

The 1936 training season was one of the least productive in years. Training was on a subdivision level, in part, no doubt, because many units were in the process of getting new equipment, but even here little of interest was apparently done. Liddell Hart reported that the First Tank Brigade had been so depleted by the loss of equipment that it "merely represents a tank brigade" and, as a result, it could only conduct company-level exercises.[58] Even more depressing was the dangerous state of mind into which the army was sinking. During the 1936 exercises, it had become increasingly apparent that attention to detail and methodical planning and movement had become so great that the army was losing much of its flexibility and mobility despite the mechanization of its equipment. The Tank Brigade, of course, had remained outside this pattern under Hobart's direction, but obviously could not remain totally aloof from this atmosphere. Liddell Hart thought this so important that he devoted a major portion of his 1936 summary of the state of the army to this problem, commenting:

> Conformity to the manuals, confusion due to constant changes in detail, and excessive emphasis on meticulous order writing have produced a slow-motion habit which is inimical to the exploitation of opportunities. Instead of building a tactical art of the basis of battle drill, thus quickening its application, we have turned tactics into a drill in slow time. The effect is seen in the attitude which can be described as "reckless caution."[59]

The one positive result from the summer's training came from the continuation of the experiments being conducted by the Third King's Own Hussars.[60] That regiment, it will be recalled, had been examining the extent to which motorized cavalry (that is, troopers mounted in trucks) could replace horsed cavalry and had concluded the year before that such formations could fulfill the cavalry's mission far more efficiently than horsemen. Now, however, the army reversed its position. Both the commander of the Third King's Own Hussars, now Lieutenant Colonel J. D. L. Wend-Fenton, and his superior, the commander of the Second Cavalry Brigade, Brigadier Heydaman, agreed that troops used in this manner were no more than infantry, and if such troops were not trained as infantry, they were not very effective. Based on his observations, Brigadier Heydaman concluded:

> I decidedly feel myself that motorized cavalry as at presently equipped are only inferior mobile infantry, but if cavalry are mechanized in light tanks, they can combine all their old roles and maintain their cavalry dash, and reconnaissance, and fire with flanking movements and shock action.[61]

On November 13, a meeting was held at the War Office to discuss the future of the mechanized cavalry. For the first time, the mechanization of the entire arm was discussed in concrete terms, and it was generally agreed that all cavalry light-tank regiments should be organized on the same pattern. No specific reference was made to motorized cavalry, but there was general agreement that some form of dismounted support was needed and that such support should be transported in cross-country tracked vehicles.[62] Shortly after, however, the idea of motorized cavalry was dropped altogether—no doubt for the reasons Wend-Fenton and Heydaman stated—and the army decided that all mechanized cavalry units would be converted into light-tank regiments. As a result, the mechanized cavalry brigades of the Mobile Division were now reorganized to include two cavalry light-tank regiments and one mechanized infantry-rifle battalion.[63] Thus ended the motorized-cavalry concept.

The fall of 1936 was a particularly depressing one for the tank advocates. The army was mechanizing at an increasingly rapid pace, but it was following a course about which they had mounting reservations. In September, Hobart wrote Liddell Hart that, because of the loss of its most modern equipment and the high turnover among its personnel, "the fact is that the Tank Brigade is not as efficient or as well trained now as it was at the end of its first season."[64] Liddell Hart, too, became depressed and considered leaving *The Times* as he was beginning to have "increasing doubt whether any progress one can help to achieve can possibly be adequate to the need and the time. In which case one's efforts may be merely encouraging a delusion."[65] In October, Hobart put his frustrations into an official report on the status of the Tank Brigade. He began by noting that, until a few years previously, the Royal Tank Corps generally had been regarded as the most advanced tank force in the world, but that this was no longer the case. Its equipment was old and worn out, and, despite promises by the War Office, replacements were not in sight. There was such a critical shortage of personnel that the brigade was forced to man tanks with incomplete crews. Among the recruits they received, moreover, there was too high a turnover rate, and Hobart cited a number of instances in which recruits were pressured to go into arms other than the Royal Tank Corps. In short, the First Tank Brigade, which had only three years earlier been a revolutionary force, was now only a relic of history.[66] Perhaps even more depressing was

Hobart's belief that his report "no doubt . . . will be suppressed when submitted and will fail to reach the eyes of anyone prepared to do anything drastic."[67]

Hobart's frustration was natural in light of the deteriorating condition of the Tank Brigade, but, ironically, it came to a head at the very time that the army was drawing up its most ambitious program for tank construction since the summer of 1918. Written by Deverell and signed by Secretary of State for War Duff Cooper, the program was submitted to the CID Sub-committee on Defense Policy and Requirements on October 19, 1936.[68] It opened with a description of the state of the army's tank forces that confirmed the substance of Hobart's report. Great Britain had a total of only 375 tanks of all types, the great majority of which were either the obsolete Vickers mediums or Carden-Lloyd machine-gun carriers used as light tanks. The prospects for any improvement in the near future were equally dismal, for, with the exception of the Mark VIA light tank, the army was not even close to the production stage of any new-model tanks. Experiments were being conducted with a new-model medium tank to replace the old Vickers medium, but even if everything went well, it would be a year before production could begin. This situation was made even more complex by the recent establishment of two new classes of tanks: infantry tanks, to fill the ranks of the army-tank battalions, and cruiser tanks, to replace the light tanks of the Tank Brigade to give it additional power. As yet, however, only one infantry tank had been produced and was undergoing tests, and the cruiser tank had not even gotten off the drawing board. Thus, the British not only were desperately short of tanks but also lacked new models with which to go into production.

In an obvious attempt to whitewash the army, Deverell placed the blame for this state of affairs on the government's refusal to provide the army with adequate funds; the disarmament conferences, which had restricted the development of heavy tanks; the reduction in arms traffic, which had led to a curtailment of weapons production; the death of several leading designers; and delays in the development of prototypes because of government procedural restrictions. Notably, he did not make any references to the army's neglect of its armored forces or to its oft-stated complacency regarding their development. These were serious omissions, for, besides distorting the events that had produced the current state of deterioration, they also laid the justification for not putting more of the tank advocates in leading positions in the future.

Nonetheless, the new proposal designed to correct this situation was by far the most comprehensive tank program the army had yet devised. The previous May, the army had established its basic tank requirements at slightly over 600 vehicles, costing approximately £12 million, but since then it had decided to mechanize the remaining cavalry regiments and develop the cruiser class of tanks to replace the light tanks in the Tank Brigade. As a result, the army now required four new types of tanks: a light tank, for the mechanized cavalry regiments; a cruiser tank, for the Tank Brigade; a new medium tank, also for the Tank Brigade; and an infantry or assault tank, for the army-tank battalions. In addition, these changes almost doubled the army's tank requirements, and Deverell now proposed the construction of slightly over 2,000 vehicles, which he believed were necessary to meet the army's immediate organizational requirements and its need for war reserves.[69] The following chart breaks down these proposals in detail:

	Light	*Medium*	*Cruiser*	*Infantry*	*Total*
May 1936 Requirements[1]	323	86	–0–	200	609
October 1936 Status	209	166	–0–	–0–	375
October 1936 Requirements[1]	584	146	132	240	1,002
October 1936 Proposal[2]	600	400	350	680	2,030

1. Includes only requirements for peacetime organization.
2. Includes war reserves.

On first glance, these figures would seem to indicate that the tank advocates had finally won their battle with the army. An extensive building program now had official sanction that would not only increase the total number of tanks fivefold, but also would give the British Army a higher proportion of tanks in relation to its size than any other army in the world. Only the Russians would have a greater actual number of tanks, but the size of their army was also far larger than the British.[70] It was an impressive sign that any earlier prejudice against tanks was dead. Deeper analysis reveals, however, that the tank advocates' victory was far from complete. The theory of armored warfare was based upon the combination of mobility and striking power[71] in sufficient quantity to enable an armored force to paralyze an enemy. In the 1936 proposal, however, the only machines capable of combining these two characteristics—the medium and cruiser tanks—accounted for merely one-fourth of the total tanks requested. Of the remaining number, 584 were light tanks, which had the mobility but lacked the striking power, and 240 were infantry tanks, which had the striking power but lacked the mobility. There could be no better proof that the army still sought to fit tanks into its traditional strategic doctrines rather than use them as a means to revolutionize those doctrines and that it continued to adhere to the "cavalry concept" of independent tank operations. The role of the cavalry in reconnaissance and screening operations would be conducted by light tanks, and its role in flank attacks and raids would be conducted by the medium cruiser tanks of the Tank Brigade. The decisive blow would still be struck by the infantry supported by the heavy infantry tanks (in place of the heavy cavalry). That the War Office still believed it was on the right track was revealed in Deverell's concluding remark that

> while certain other countries, notably Russia, may be ahead of us as regards the design of certain types of tank, it is probably true to say that we are ahead of other nations in training and tactical handling.[72]

On October 19, Cooper submitted this request to the Treasury Inter-Service Committee, and it began a long and excruciatingly slow struggle through the government bureaucracy. A more immediate problem faced the tank program, however, and its solution was no less slow in coming: the technical problems of the design and production of the vehicles themselves. In theory, the responsibility for research and weapons development rested with the MGO, who was supposed to be guided on broad policy by the Army Council as a whole and in more detail by the CIGS's Specifications Committee. There was, however, little harmony between the departments under the CIGS and MGO. The General Staff argued that policy could not be established until a prototype tank was available for trial, but the Directorate of Mechanization (under the MGO) maintained it could not develop prototypes until the General Staff had provided some policy guidance.[73] This difficulty was compounded by the army's shortage of qualified technical officers and the disregard with which they were treated by the other arms. This situation was the product of the generally low status given to technical officers, who seldom came from the country's social elite, and the British colonial tradition, which emphasized the value of the "commonsense" officer who could accomplish his mission on a shoestring.[74] As a result, the tanks developed during this period were largely the product of the firm of Vickers-Armstrong, Ltd. Their design section was far stronger than that of the War Office, and the army generally accepted this situation as quite satisfactory.[75]

These problems were matched by equally serious difficulties in production. Disarmament and public pressure against the arms trade had so reduced the market for armaments that, of the four great armaments firms in 1913, only Vickers-Armstrong remained in 1934. It was supplemented by three Royal Ordnance factories, which in 1933 only employed one-tenth of the people they had in 1918.[76] These conditions naturally affected all aspects of the British rearmament program, but they were especially detrimental to tank development, as tanks had no civilian courterpart on which to rely for general technical advances or productive capacity. Victor Germains had noted this problem in 1928, when he declared that Fuller and Liddell Hart's view of the relationship between war and civilization did not apply to tracked vehicles, and he was proven right by events. If tanks were to be produced, their development would have to be underwritten by the army and after almost twenty years of neglect, it would take time to build new machinery. Unfortunately, the War Office's program calling for four different types of tanks—a direct consequence of the cavalry concept of tank warfare—added to the problem by spreading the army's already thin resources in this area.

The 1937 army estimates reflected both the army's plans and problems. The army was allotted a total of £63 million, the second year of significant increases, and, as in the previous year, most of it was devoted to reequipment of the older arms. A total of £6.3 million was to be devoted to mechanization, which would enable the mechanization of the artillery and signals to be completed and that of the infantry to be largely done. Little could be expected in the way of large numbers of tanks, however. Although the tank advocates were obviously critical of this decision, the army actually had little choice in the matter as, with the exception of the Mark VIA light tank, it had no model with which to go into production.[77]

The Loss of British Tank Supremacy

Although the British Army could justly point to a number of advances achieved during this period, these successes were overshadowed by the fact that it had lost its position as the world's foremost tank power. The cause was not so much the decline of its own forces as it was the rise of a new power: Nazi Germany. The Germans had formed their first tank brigade only in late 1934, but a year later they expanded it into three Panzer divisions. These advances in substance were matched by similar advances in theory. Inspired in part by Fuller and Liddell Hart, a small group of German officers led by Heinz Guderian were, with the support of Hitler, able to override the more conservative elements in the German Army and firmly implant the new theories in its strategic and tactical doctrines. For the first time since 1918, England faced a serious threat from a major power, a power that was building its army around the offensive power of the tank.

1. Sir Henry Pownall, *Chief of Staff: The Diaries of Lieutenant General Sir Henry Pownall,* vol. 1:*1933–1940,* ed. Brian Bond (London: Archon Books, 1973), pp. 39 and 43.

2. WO 32/2847, "Introduction of the Mobile Division," Minute 1, dated 15 October 1934. It is equally significant that Montgomery-Massingberd still felt Lindsay was the best possible choice to command the new unit, but the division was not actually formed until 1937, when a new CIGS was in office. Lindsay was sent to India, where he served until his retirement in 1939. Basil H. Liddell Hart, *The Tanks,* 2 vols. (New York: Frederick A. Praeger, 1959), 1:337. See also letter by Hobart to Lindsay, October 18, 1934, in which he wrote that the CIGS was favorable to this appointment. File 3, Major General John Lindsay Papers. Liddell Hart Archives, States House, Medmenham, Marlow Buckinghamshire, England.

3. Colonel G. N. MacReady, "The Trend of Organization in the Army," *Journal of the Royal United Service Institution* 80 (February 1935): 14.

4. WO 33/1509, "Report by a Committee Assembled to Consider the Organization of the Mechanized Cavalry and the Royal Tank Corps 1938," pp. 12–13; WO 32/2841, "Cavalry

Committee Interim Report, 1935," pp. 5–7; Lieutenant Colonel Maitland-Dougall, "The Future of Land Warfare," *Army Quarterly* 27 (January 1936): 262–69.

5. See above, pp. 109–10.

6. *Daily Telegraph,* 12 April 1928.

7. B. H. Liddell Hart, "Armoured Forces in 1928," *Journal of the Royal United Service Institution* 73 (November 1928): 728.

8. Charles Broad to Liddell Hart, November 23, 1928, Liddell Hart Papers.

9. "The Organization of Tactical Vehicles," an unpublished paper in Hobart Papers, Liddell Hart Archives, State House, Medmenham, Marlow, Buckinghamshire. The original draft was written by Hobart in late 1934 and sent to Lindsay on January 17, 1935, with corrections following on January 28. On February 4, Lindsay returned both with comments of his own. The final draft was later incorporated into a larger paper entitled "Memorandum on the Organization of Tanks and Mobile Troops" and submitted to the DSD on 22 March 1935.

10. Liddell Hart, *The Times,* 27 December 1935.

11. WO 33/1509, pp. 11–12.

12. Ibid.

13. Liddell Hart, *The Tanks,* 1:342.

14. WO 33/1509, pp. 12–13.

15. *Daily Telegraph,* 15 November 1934. The designation "army tank battalion" was assigned to units employed in the role of infantry support to distinguish them from "mixed battalions," which formed the strike force of the Tank Brigade.

16. WO 32/2847, "Introduction of the Mobile Division, 1934–1936," Minute 1. The "Western Plan" was the code name for a Continental commitment. The details of the Expeditionary Force for use in Europe were spelled out in WO 33/1341, "Composition of the Field Force and Remaining Formations and Units Not Alloted to the Field Force in the Western Plan, 1934." See also WO 32/4612, "Future Organization of the British Army, 1935," p. 9.

17. WO 32/4612, "Future Organization of the British Army, 1935," p. 1.

18. Ibid., p. 16.

19. Ibid., pp. 24–25.

20. Ibid., pp. 14–15. He felt that financial limitations precluded the production of armored fighting vehicles in sufficient quantity prior to the outbreak of hostilities, and, besides, the constant advances in technology would make such a policy wasteful.

21. Ibid., pp. 4–6.

22. Ibid., p. 4.

23. Ibid., p. 20.

24. WO 32/2826, "The Mobile Division and Mechanization of the Cavalry," Minutes 1A, 2, 3, 5, and 6.

25. Ibid., Minute 1A.

26. "Army Training Memorandum Number 13," dated December 1934, Appendix A.

27. WO 32/2847, "Introduction of the Mobile Division, 1934–1936," Minute 7A.

28. *The Times,* 14 and 18 March 1935. Liddell Hart moved from the *Daily Telegraph* to *The Times* on 14 March. For criticism of this allocation of funds, see Pile to Liddell Hart, 17 September and 25 October 1935, Liddell Hart Papers.

29. *The Times,* 8 August 1935.

30. Liddell Hart, *The Tanks,* 1:339; "Army Training Memorandum Number 13," dated December 1934, pp. 12–13, 23.

31. *The Times,* 23 and 27 August 1935.

32. *The Times,* 18 and 19 September 1935.

33. WO 32/2847, "Introduction of the Mobile Division, 1934–1936," Minute 7A, dated 8 December 1934.

34. WO 32/2847, "Report of the 3rd King's Own Hussars on the Mechanization Experiment, 1935," Minute 13A, dated 25 September 1935.

35. Ibid.

36. See comments by Commanding Officer, Second Cavalry Brigade, dated 30 September 1935 and General Officer Commanding, Southern Command, dated 28 October 1935 in WO 32/2847.

37. This oversight did not pass by the tank advocates unnoticed. Pile wrote to Liddell Hart, "Of course the real truth is an attempt to save regiments that ought to have been disbanded

twenty years ago. It is the infantry who should be put into busses." Pile to Liddell Hart, 25 October 1935, Liddell Hart Papers.

38. WO 32/2826, In Minute 1, dated 15 October 1935, Montgomery-Massingberd refers to "our recent decision" to form the Mobile Division, but the actual record of that decision cannot be found.

39. Montgomery-Massingberd to Lord Halifax, 17 August 1935, Montgomery-Massingberd Papers.

40. WO 32/2826, "The Mobile Division and Mechanization of the Cavalry," Minute 1A, p. 1. See chart, p. 182.

41. Ibid., Minute 6.

42. Ibid., Minutes 2, 3, and 5.

43. Great Britain, War Office, *Field Service Regulations,* vol. 2 *Operations General* (London: HMSO, 1935), p. 4.

44. Ibid.

45. Great Britain, War Office, *Field Service Regulations,* vol. 3: *Higher Operations* (London HMSO, 1935), p. 24.

46. Ibid., p. 37.

47. Ibid., p. 46.

48. Montgomery-Massingberd, *Autobiography of a Gunner,* a privately published memoir in the possession of the author.

49. Montgomery-Massingberd, "Handing Over Notes," an unpublished memorandum dated 23 February 1936. Montgomery-Massingberd Papers.

50. Montgomery-Massingberd, *Autobiography of a Gunner.*

51. *The Times,* 15 March 1936.

52. On 10 May 1940, the Germans had a total of 136 divisions, of which ten were Panzer and six were motorized infantry. This compares with a total of five British divisions in the prewar army. Larry H. Addington, *The Blitzkrieg Era and the German General Staff, 1865–1941* (New Brunswick, N.J.: Rutgers University Press, 1971), pp. 97–98.

53. General Sir Frederick Pile, personal interview with the author at the Cavalry Club, London, on 6 August 1971.

54. This emphasis on the army's general rearmament policy caused Liddell Hart to fail to see the differences between Montgomery-Massingberd and Deverell in later years. Writing in 1959, he commented, "Those who were now in the late 1930's in charge of the British Army had no longer as clear a vision of future warfare as their predecessors had in 1927, and missed the opportunity of developing the new arm in their concern to maintain and renovate the older arm." *The Tanks,* 1:355. Such a view is not only unfair to Montgomery-Massingberd, but also reflects an excessively narrow view of mechanization and the requirements of the army.

55. Robin Higham, *Armed Forces in Peacetime: Britain, 1918–1940* (Hamden, Conn.: Anchor Books, 1962), pp. 325–27. The army estimates for the year came to £55 million.

56. Liddell Hart, *The Tanks,* 1:359–60.

57. Hobart to Liddell Hart, 30 May 1936, Liddell Hart Papers.

58. *The Times,* 3 and 4 September 1936.

59. *The Times,* 2 November 1936. An Army Training Memorandum made the same point the following year. It stated, "There is a general tendency throughout the Army for operations to become unduly slow and methodical." "Army Training Memorandum Number 18," dated April 1937.

60. WO 32/2847, "Introduction of the Mobile Division, 1934–1936," Minutes not numbered.

61. Ibid., "Report of the 2nd Cavalry Brigade on the 3rd King's Hussars Experiment," Minute 37, dated 13 October 1936.

62. Ibid., Minute 40A, meeting of 13 November 1936.

63. "Army Training Memorandum Number 17," dated January 1937, p. 30.

64. Hobart to Liddell Hart, 21 September 1936, Liddell Hart Papers.

65. Liddell Hart to Hobart, 20 September 1936, Liddell Hart Papers.

66. Hobart, "Report on the Status of the Tank Brigade," an unpublished memorandum. Hobart Papers.

67. Hobart to Liddell Hart, 10 October 1936, Liddell Hart Papers.

68. WO 32/4441, "The Tank Situation, 1936–1940," Minute 5B; "The Tank Situation, 1936," pages not numbered.

69. Ibid., "W. O. Memorandum 57," Minute 12A, dated 10 November 1936. WO 32/4441. The May requirements were contained in WO Memorandum 9. This memorandum is not in WO 32/4441, but is referred to in WO Memorandum 57 and in the Minutes of the 13 November 1936 meeting of the Treasury Inter-Service Committee, both of which are in the file. There is no indication that any action was taken as a result of WO Memorandum 9 by the War Office or the Treasury Inter-Service Committee. As the army failed to produce anything approaching a satisfactory model in any of their classes of tanks between May and October, the Memorandum undoubtedly remained a dead letter and was superceded by WO Memorandum 57.

70. Ibid., "The Tank Situation," Minute 5B. The army estimated that the United States had thirty-five light and fifteen medium tanks, Germany had 1,600 light and 400 medium, France had 600 light and eighteen medium and Russia roughtly 2,000 light and 4,000 medium tanks.

71. Striking power was defined as the combination of firepower and protection.

72. WO 32/4441, "The Tank Situation, 1936," Minute 5B.

73. M. M. Postan, D. Hay and J. D. Scott, *The Design and Development of Weapons* in *History of the Second World War, United Kingdom Civil Series,* Sir Keith Hancock, gen. ed. (London: HMSO, 1964), p. 306.

74. Martel noted this in a letter to Liddell Hart on April 13, 1928, Liddell Hart Papers; and the British Official History of the Second World War matter-of-factly states: "Within the Army the technical officer was generally despised." Postan, et al., *Design and Development of Weapons,* 240.

75. Postan, et al., *Design and Development of Weapons,* pp. 244, 306.

76. M. M. Postan, *British War Production* in *History of the Second World War, United Kingdom Civil Series,* Sir Keith Hancock, gen. ed. (London: HMSO, 1952), p. 8.

77. *The Times,* 5 March 1937; Liddell Hart, *The Tanks,* 1:384; Higham, *Armed Forces in Peacetime,* pp. 326–27.

• 7 •
Prelude to War: 1937–40

Introduction

IN the three years prior to the battle of France, the British Army sought desperately to prepare itself for what it was certain was the coming confrontation with Germany. Although it was still hampered by a government policy that gave it the lowest priority among the services, the army nonetheless was able to begin rebuilding its strength for the first time since the end of the war. For the armored forces, this meant rapid expansion and the influx of new equipment on a scale undreamed of only a few years earlier. As a result, the stakes in the struggle for control of these burgeoning forces assumed new and ominous proportions. No longer did the army or the advocates of the strategy of armored war attempt to convert each other to their way of thinking; instead, they engaged in a naked struggle for control of the crucial command and staff positions within the army that would determine its future development. It began with the appointment of a new secretary of state for war in the spring of 1937.

"The Partnership"

On May 28, 1937, Neville Chamberlain succeeded Stanley Baldwin as prime minister and shortly thereafter transferred Duff Cooper, who had been secretary of state for war, to the Admiralty, replacing him with a young and well-known politician by the name of Leslie Hore-Belisha. Only forty-one at the time of this appointment, Hore-Belisha had served in the Royal Army Service Corps in World War I and in Parliament since 1923. He first had entered a government ministry in a minor capacity in 1931 and three years later was appointed minister of transport, earning the reputation of an aggressive and dynamic reformer. His new appointment was seen as a definite indication that the government was determined to carry out widespread reform and was greeted, in some quarter at least, with high hopes.[1] Hore-Belisha suffered from two handicaps, however, that severely curtailed his effectiveness. The first was anti-Semitism. Cooper had warned him this might be a problem when Hore-Belisha assumed his new post, and, although it is difficult to substantiate positively or measure precisely, there are indications that it existed.[2] His second handicap, however, was probably more serious and was self-inflicted: it was an almost complete ignorance of military affairs. Except for his wartime service, he had had no contact with the army, and there is no indication that he had made any effort to keep up with the current military problems. Suprisingly, this lack of knowledge did not bother him; as he explained to Liddell Hart shortly after taking office, he had known nothing of the Treasury when he went there in 1932, but had found he could reason things out logically and reach sound decisions nonetheless.[3] This approach proved to work fairly well when dealing with problems such as the welfare of troops and army administration, but it soon became evident that more specific knowledge was necessary to grasp the more complex problems facing the army.

It was undoubtedly out of this lack of knowledge that "the Partnership" of Hore-Belisha and Liddell Hart formed. They first met on June 7 at a lunch arranged by Duff Cooper at which they discussed some of the problems facing the army. At Hore-Belisha's request, Liddell Hart afterwards sent him a proposal for army reorganization on which he had been working. The two did not meet again until August 18, when they agreed Liddell Hart would act as the secretary's unofficial advisor.[4] For the next ten months, they were in almost constant contact. They saw or spoke to each other almost daily and, during periods of crisis, as often as several times a day. Given this combination—a dedicated reformer and a leading proponent of armored warfare—it might be expected that Great Britain's armored forces and the theory of armored war would have emerged boldly from the shadows in which they had existed for so long. But they did not.

The reasons for this failure are difficult to pinpoint. In his memoirs, Liddell Hart laid the blame on the influence of conservatives in the War Office. Until December 1937, one of the leading obstructionists was Deverell, but his removal did not fully solve the problem, and, having purged the Army Council once, Hore-Belisha felt he could not do so again.[5] Much truth as there is to this, it was not the only reason. At least as important was Hore-Belisha's apparent lack of understanding of the theory of armored warfare. Although the secretary received numerous memoranda and proposals from Liddell Hart predicated on this strategy and gave the appearance of accepting and trying to act on it, the conversations between the two men indicate he had little real understanding of what was required. He was, for example, all too willing to seek a compromise commander for the Mobile Division and had little comprehension of why his "partner" insisted that armor development be placed in the hands of the tank experts. Consequently, despite Liddell Hart's urging, many necessary tasks were left undone.

The 1937 army exercises demonstrated many of the problems "the Partnership" faced. The theme for the training was "the problem of the offensive battle" with special emphasis on tank-infantry attacks, night movement and attacks, and maintaining the momentum of the attack. All of the exercises were on the brigade and division level, which meant they were of little value to the senior commanders, and not only was the role of tanks less than it had been in years, but also their use was far too slow and methodical.[6] A War Office skeleton exercise held in September provided an even more disturbing indication of the army's state of mind. The exercise called for a British force of four infantry divisions and the Mobile Division to move against an enemy force through a corridor forty miles wide. The commander of the British force, General Sir Edmund Ironside, planned to advance on a broad front to pin the enemy forces down and then send the Mobile Division on a sweep around the enemy's flank to strike at his rear. The enemy withdrew before the attack could take place, however, and Ironside merely followed with a conventional advance. He reached the height of absurdity when he placed the Mobile Division on the line and attempted a flanking maneuver with one of the infantry divisions. Although both Ironside and the Mobile Division commander were severely criticized for their conduct of the exercise, the forty-mile restriction was no less absurd in light of the proven mobility of tank forces, and once again demonstrated the highly restricted view of the tank still prevalent among senior officers.[7] It was a point well made by Hobart that summer when he told Liddell Hart,

> It is not, I think, at all fully realized yet that the mechanization of the Army is not only a matter of supply of vehicles and training in their use, but demands a fundamental *mental* change as great as the change in the Navy from wooden sailing ships to steamer ironclads.[8]

Another source of growing frustration to the tank advocates was the extremely slow pace at which the army's tank strength was increasing. The army's proposal of October 1936 had been submitted to the Defense Policy and Requirements Subcommittee of the Committee of Imperial Defense on November 10 and received the full support of that body when it met three days later. The committee—which, significantly, included Deverell—agreed there was an urgent need for more rapid tank development and recommended that the army go beyond the firm of Vickers, Ltd., for new tank designs and establish closer contacts with the industry.[9] It then passed the proposal on to the Treasury Inter-Service Committee, which took up the matter on November 23, which, in turn, forwarded it to the chancellor of the Exchequer.[10] At this point, the tank program came to a grinding halt while the Exchequer and the Treasury Inter-Service Committee passed the request back and forth for further information and clarification of details. In general, their questions revolved around the increases in the estimates caused by the decision to mechanize additional cavalry units, the revisions in tank specifications, which added to their cost, and the problem of determining what constituted an adequate reserve to make up for immediate losses in the event of war. The situation was further complicated when the army requested the addition of machine-gun carriers for the mechanized cavalry regiments and further modifications of the vehicles requested, which brought the total cost of the program to £29,741,000 as compared to the £23,771,500 requested in November.[11] On March 23, 1937, the Treasury Inter-Service Committee met to consider the proposal as it then stood. It refused to approve the entire program on the grounds that such a large expenditure could only be authorized by the chancellor of the Exchequer, but it did give the army permission to order the machine-gun carriers for the mechanized cavalry and the additional light tanks it had requested. In both cases, the items approved called for relatively small expenditures and the army had models with which it could go into production.

Such was not the case with the medium, cruiser, and infantry tanks the army had requested, for all three were plagued with design problems. Martel, who was then assistant director of mechanization, described the situation in a letter to Liddell Hart in August 1937. He wrote that medium-tank design had been in a chaotic state since the failure to adopt the "16 tonner" in 1931, and the only model then in existence was underpowered and underarmored. He expressed the hope, however, that a new design nearing completion would prove satisfactory. As for the cruiser and infantry tanks, there were still no models capable of meeting the army's specification, but once again he had hopes for new designs that were then being developed.[12] In short, the army had made no concrete advances in developing satisfactory models of cruiser and infantry tanks since it had submitted its original proposal the previous October. The War Office itself admitted this when it informed the Treasury Inter-Service Committee that "the production of actual tanks is not held up at present by the absence of such authority [to order them from manufacturers]."[13]

In July, Hore-Belisha opened another effort to get the tank program moving. In an internal War Office memorandum dated July 23, he placed special emphasis on the new tanks for the Mobile Division, writing: "So far as I am concerned the most urgent matter is to get authority for the Cruiser Tanks, in which there is a reasonable probability of getting into production and for which production can be prepared now, eg. engines."[14] Three days later, the War Office submitted another request for authority to order medium, cruiser, and infantry tanks.[15] Noting that the total program of defense requirements, including funds for mechanization, had been approved by the government, it requested that the Treasury Inter-Service Committee approve the request as soon as possible so that production of certain components could begin immediately to avoid bottlenecks when the vehicles went into production later. In an obvious attempt to placate the cost-conscious Treasury, the War Office further noted it was requesting sufficient funds to meet the army's organizational needs only and not for the additional machines to cover war reserves.[16] On July 27, the Treasury Inter-Service Committee met and agreed to allow the army to place the following order: 50 medium tanks, the components for 548 infantry tanks, 164 medium tanks and 271 cruiser tanks, and the subsequent construction of the infantry and cruiser tanks, and the guns and ammunition for an additional 83 medium tanks, 285 infantry tanks and 79 cruiser tanks partially to meet war-reserve requirements.[17] On October 16, the Treasury gave its approval to this program plus an additional 360 light tanks to those authorized earlier and 1,617 machine-gun carriers the Treasury Inter-Service Committee had authorized earlier. The question of war reserves, the Treasury noted, would not be contemplated until the possibility of holding such reserves in the form of manufacturing capacity could be studied.[18] In any event, the army's tank strength, when compared to that proposed in October 1936, had increased by fifty percent as the following table indicates:

	Light	*Medium*	*Cruiser*	*Infantry*	*Total*
October 1936[1]	584	146	132	240	1,102
October 1937[1]	891	164	271	548	1,874
October 1936[2]	600	400	350	680	2,030
October 1937[2]	1,327	247	350	833	2,757

1. Peacetime strength only.
2. Peacetime strength plus war reserves (not authorized at this time).

Although these figures make an impressive presentation and demonstrate that the army was anything but "antitank"—remember that these figures represent the tank support for a peacetime army that contained only one mobile and four infantry divisions at home and scattered overseas garrisons—it still did not indicate any change in the army's strategic doctrines. As in the proposal of the previous October, over half of the new tanks were either classified as "light" or "infantry" and would thus be of little value in long-range operations against enemy objectives. More important in immediate terms, however, was the fact that, except for the light tanks, the army had yet to find satisfactory models for any of the other classes.

Under these conditions, it is obvious why the tank advocates looked directly to Hore-Belisha for help in their cause and why Liddell Hart, once be became the secretary's unofficial advisor, began to urge strongly that they be placed in key positions to control the future course of mechanization. The situation had been so terribly mishandled by the War Office that they found it impossible to believe that one individual could correct it by himself. As Pile wrote to Liddell Hart in June, "He [Hore-Belisha] may get Deverell and his satillites to agree to such a reorganization but all the little difficulties that are bound to crop up will so retard the progress of the machine that H.-B. will be with his ancestors long ere he can emulate Haldane."[19]

Liddell Hart opened his campaign to place the leading tank advocates in key army positions only a few days after his renewed contact with Hore-Belisha in August. In a specially prepared memorandum on this subject, he wrote:

> To gain light on the problem of mechanization and the troubles which have made progress so slow, the best way is to gather the views of the men who have been connected with its evolution and have been exploring these problems for years past. The greatest difficulty today is that the Army had been suddenly transformed to a mechanized and motorized basis so recently that hardly anyone in the higher posts has any personal knowledge of its practical side. During the experimental years that [experience] has lain with junior officers. A sprinkling of the best ones are now at last rising to somewhere near the top, if not yet high enough to have direct access. Of these, Broad and Pile have recently become Major Generals, Hobart is near the head, and Martel well up, the colonels list.[20]

Liddell Hart did not let the matter rest here. On the day he presented this paper to Hore-Belisha, he took the opportunity to stress the need for an MGO who was well-versed in both artillery and mechanization problems, and noted that both Broad and Pile, as artillerymen with extensive mechanization experience, were two of the few qualified men in both fields. He returned to the subject on September 4, when he again noted the lack of tank experience among the commanders of the regular infantry divisions. He admitted that the choices were limited, but felt this omission went beyond the matter of availability:

> It is not that the people now at the top fail to appreciate the merits of the leading Tank Corps officers—they speak most highly of them. But when it comes to filling key posts of a general kind they seem frightened lest these men go too fast in modernizing ideas, training, and organization. Yet in belatedly adopting the mechanization of the army, with all the needless hustle of the present programme, the chiefs of the army have tacitly admitted that such men were sure-sighted when they themselves were shortsighted.[21]

On September 7, he again stressed the need to have the right man in key War Office positions and four days later urged Hore-Belisha to appoint Broad or Hobart as commandant of the Staff College (where they could ensure rising officers would be properly trained in modern war) and Pile as MGO or commander of the Mobile Division.[22] This campaign to have the leading tank advocates placed in key positions clearly demonstrates that Liddell Hart no longer felt the officers of the older arms were capable of conversion to the new warfare on their own. Although he had known all of the tank advocates for many years, these recommendations were not in his eyes based on favoritism. By this time, he had served in or been observing the army for twenty years; he knew virtually all of the senior officers and had made many attempts to win them over to his views. It was their failure to recognize the value of his ideas combined with his own certainty that he was correct, rather than any favoritism, that prompted him to recommend the men he did.

What particularly disturbed the tank advocates was, of course, the dominance of the cavalry concept of the role of the tank and how this was affecting not only the role of the Mobile Division, but also its organization and equipment. In October, Liddell Hart had a particularly disturbing conversation with Burnett-Stuart, then general officer commanding, Southern Command, in which the latter expressed his conviction that the replacement of the cavalry division with the Mobile Division was merely a change in transportation, not purpose, and that the role of both units should be exactly the same.[23]

The attitudes expressed in the *Royal Tank Corps Journal* must have been even more disturbing to the advocates of armored war. From 1937 on, the journal naturally concentrated heavily on the role of tanks in future wars, both as direct infantry-support weapons and as highly mobile formations to strike at the enemy's flanks and rear areas. The authors—whether from the infantry or the Royal Tank Corps itself—never questioned the absolute necessity of tanks for the successful prosecution of the offense in either role: but neither did they question the concept of the decisive battle or, by implication, the strategy of attrition. One infantryman, who was asked to write an article by the journal's editor, claimed that the Tank Corps had "betrayed" the infantry by relegating the tanks' direct support role to a minor function. Building on the "slender thread" of a few isolated examples of tanks that had penetrated into German rear areas during World War I, he claimed that the corps had reduced itself to a mere "glorified cavalry" and ignored its great traditions.[24] Another infantryman admitted that "the purely infantry assault, already obsolete in 1917, is almost too fantastic to mention," and that "against a co-ordinated field army defense assaulting power must be rated in terms of armoured fighting vehicles." Yet he, too, claimed that the efforts to develop mobile formations for strikes against an enemy's rear areas "are without avail unless we also possess a main battle force capable of fixing and exhausting the hostile defensive power, which would otherwise counter the action of the mobile arm."[25] Other authors, presumably from the Royal Tank Corps, emphasized the role of mobile armored formations to a far greater degree. Their vision went beyond that of the infantrymen just noted, and they often stressed the value of strikes deep into the enemy rear aimed at his command structure. Yet they, too, failed to see the paralyzing potential of such strikes or that such actions could eliminate battles in the traditional sense. One, in fact, while noting that such strikes were more than mere raids, nonetheless stated that "to be really effective the action of a Tank Brigade must result in the distraction of the enemy H. Q., causing divided attention, nervousness about their rearward services and consequent division of troops from the main battle."[26] Clearly, the advocates of the theory of armored war were not simply a minority in the army as a whole; they were also a minority in the Tank Corps itself!

As long as such views remained dominant, the mission of the Mobile Division would remain primarily that of reconnaissance and pursuit, its training would omit offensive operations and deep, strategic penetrations, and it would contain an extremely large number of light tanks, which, while perfectly adequate for the cavalry role, were totally inadequate for offensive action. This uneasiness was brought out in two papers—one by Hobart and the other by Liddell Hart—both written in October 1937. Hobart's theme was that the Mobile Division as then constituted combined the two roles of armored formations, reconnaissance and offensive action, and that this was a fundamental weakness, as each role required different organizations and equipment. To correct the situation, he proposed the formation of two types of mobile division: a light-tank division of two mechanized cavalry brigades, each with three light-tank regiments for strategic reconnaissance and protection; and a tank division composed of one cavalry brigade of two or three light-tank regiments and a tank brigade. He also suggested that two tank divisions be formed to add to the possibilities of deception and surprise.[27] Hobart's paper was obviously unacceptable to the War Office, and he was, in fact, told "that I was completely obsessed by tanks and had quite lost sight of the Field Force in general and the nature of the operations in which it might be engaged."[28]

Shortly after hearing of the War Office's rejection of Hobart's paper, Liddell Hart attempted to correct the situation by going directly to Hore-Belisha. On October 31, he sent the secretary his own paper in which he traced the history of the development of the Mobile Division and described its effectiveness for offensive action in a European war. The paper was not an exposition of armored war (which Hore-Belisha could have used), but was clearly predicated on those ideas. Its main themes were the need for an armored force capable of offensive action and the fact that such a force would require heavier tanks than those with which the Mobile Division would be equipped.[29] Unfortunately, nothing came of the paper. There are no references to it in any later conversations between "the Partners," and, if it was ever circulated in the War Office, it never made any impression. The army continued to separate mobility and offensive power to the detriment of both.

In the meantime, Hore-Belisha moved on his own to change the War Office by appointing a new military secretary. The primary function of this official was to advise the secretary of state for war on appointments, and he was thus one of the more important figures in the army. Hore-Belisha's decision to appoint a new secretary stemmed from his "desire to appoint one of the younger Major-Generals as an indication to the Army that we wish to encourage promotion at as early a stage as ability and efficiency justify."[30] His choice was Major General Lord Gort, then commandant of the Staff College, who was only forty-nine years old. The proposal immediately aroused the strong opposition of Deverell, who argued that Gort would be better as a division commander, but beneath the surface there were indications that Deverell was more concerned that the appointment did not follow the rules of seniority.[31]

The conflict over Gort's appointment was only a small skirmish compared to what followed, for Hore-Belisha and Deverell disagreed at almost every turn. The fact that the secretary was temperamentally incapable of being a mere figurehead was a fundamental source of irritation to the military members of the Army Council, who felt he was interfering too much with their authority, but the conflict quickly centered on questions of policy. The appointment of younger officers to senior positions and the course (but not the fact) of mechanization were two other areas of conflict, but they were by no means the only ones nor were they apparently the most important. Improvements in barracks, better rates of pay, and the provision of many new services for enlisted men were all matters in which Hore-Belisha exhibited a great deal of interest that was hardly matched by the Army Council.[32] Most important, however, was the army's opposition to the policy of "limited liability" and to the influence of Liddell Hart, who was perceived to be one of the chief architects of this policy. The War Office had never altered its belief that the army should be prepared to fight on the Continent, and, when it appeared this was not to be, it reacted with enraged frustration. As General Pownall confided to his diary:

> They (the Politicians) cannot or will not realize that if war with Germany comes again (whether by collective security, Locarno or any other way) we shall again be *fighting for our lives.* Our effort *must* be the maximum, by land, sea and air. We cannot say our contribution is "so and so"—and no more, because we cannot lose the war without extinction of the Empire. The idea of the "half-hearted" war is the most pernicious and dangerous in the world. It will be 100 percent—and even then we may lose it. We shall certainly lose it if we don't go 100 percent. In God's name let us recognize that from the outset—and by that I mean *now.*[33]

When Hore-Belisha made it clear that he not only supported "limited-liability," but would even judge an officer's fitness for a War Office appointment on this basis, the army's attitude toward him was settled.[34] He could do no right. Unfortunately for all concerned, Liddell Hart was placed in the same category for the very same reason despite the fact that he differed with the government on certain fundamental aspects of how this policy should be applied. The government's position was based on budgetary considerations and an emotional reaction to the Western Front. To them, "limited liability" meant the army would not be employed on the Continent and therefore its modernization should be given the lowest priority. To Liddell Hart, it meant that, while primary emphasis should be given to the navy and air force, it was equally essential to establish a small but completely modernized, highly mobile army capable of intervening on the Continent at the critical juncture. Whether or not such a force could have played the role Liddell Hart expected of it in the face of a massive German onslought is, of course, problematic. Even a German advance to approximately the same line they had reached in 1914 would have caused far greater difficulties for Great Britain now because of the advent of air power, a fact the army perceived much more clearly than he. Nonetheless, whatever the weaknesses of the policy of "limited liability," starving the army of needed equipment was not, in Liddell Hart's plan, one of them. The army, however, never perceived this fact. Pownall, again typifying their views, commented in his diary that "the writings of the ineffable Liddell Hart have done great harm. He has preached the easier course, the cheaper course—naturally it is pounced on by such people [as Hore-Belisha and other political leaders]." Some months later he further commented on Liddell Hart: "He needs to be shot dead, a sinister influence who has done the army much harm."[35]

The effect of this conflict over "limited liability" was both to greatly complicate and to obscure the differences between Liddell Hart and the War Office over the future of the armored forces. As far as Liddell Hart was concerned, it was another example of the army conservatism he had been combating for some fifteen years. To the army, however, the issue of the tank forces seems to have been completely ignored. Deverell, writing to a friend after the war, claimed he was forced to retire because of his oppositin to "limited liability" and Pownall never mentioned anything but this policy in his attacks on either of "the Partners." It would be too much to claim that the absence of the "limited liability" controversy would have enabled Liddell Hart to win the army over to his view of armored warfare at this late date, but there can be no question that it greatly limited his effectiveness by making the army far less receptive to anything he had to say.

By the late summer of 1937, relations between Hore-Bleisha and Deverell were so poisoned that both men occasionally exhibited more concern with each other's activities than with what the army was doing. On September 10, Hore-Belisha told Liddell Hart that he felt attempts were being made to cut him off from opinions except those of the CIGS and other senior War Office officials. Deverell was in no better state of mind. He believed Hore-Belisha was deliberately undermining his position, and as Liddell Hart noted at the time, he "has been looking very harassed these last few days, and also becoming as suspicious as Milne was. . . . He keeps on murmuring that there must be 'no radical changes.' Has seized many chances to talk to me, while angling to find out how much I see of Hore-Belisha."[36]

An explosion between the two could not be long averted, and it finally erupted in early September, when the CIGS recommended General Michael Blakiston-Houston for command of the Mobile Division. Blakiston-Houston, who was then commandant of the Equitation School and inspector of cavalry, had no previous experience with mechanized units of any type, and his nomination aroused the immediate opposition of the tank advocates. On September 12, Hore-Belisha told Liddell Hart he had just met Balkiston-Houston "and had not been impressed. A typical cavalry man. He [Hore-Belisha] was disinclined to give him the Mobile Division and on the whole most inclined to Broad."[37] It was soon clear, however, that Deverell was not about to take Hore-Belisha's recommendation and furthermore very much resented what he considered to be the secretary's undue intervention in army affairs.[38]

The controversy over this appointment continued for almost two months and brought the War Office to an imposssible impasse. Deverell became increasingly intransigent in his defense of Blakiston-Houston, and at one point he declared that he was the only man in the entire army qualified to command the new division. When Hore-Belisha asked what would happen if he dropped dead, he received no reply.[39] Liddell Hart, meanwhile, privately—and insistently—urged Hore-Belisha to appoint one of the tank advocates to the command. On September 12, he recommended that, if training for peace were the first criterion, it should go to Broad, and, if command in war were the prime criterion, it should go to Pile. Above both, however, he placed Hobart, who, although only a brigadier, he considered the best qualified under any circumstances. Hore-Belisha replied that he was hesitant to propose any of these names before Deverell brought them up first, as he feared it might give rise to charges of favoritism.[40] On October 19, however he mentioned to Deverell the possibility of appointing one of them on the excuse that he had met them on recent training exercises and had been impressed. His suggestion was firmly rejected by Dererell on the grounds that Broad and Pile were only secondrate, that Pile had never commanded a mechanized unit, and that, while Hobart was a gifted commander, he had been divorced, a fact that would make "the whole of the cavalry" resign if he received the appointment. More disturbing than the absurdity of these charges was that Hore-Belisha apparently could not answer Deverell on the spot and had to call Liddell Hart that evening to get a response. As Liddell Hart pointed out in their conversation, the personal files of Broad and Pile would show their high quality and also that Pile had seen considerable service with tanks. Why Hore-Belisha had not checked their records himself before this is unknown; his failure scarcely redounds to his credit.[41]

On October 31, a way out of the impasse emerged. Deverell agreed to withdraw nomination of Blakiston-Houston and replace it with Major General Alan Brooke, then DMT. When told of this, Liddell Hart expressed admiration for Brooke, but also expressed the view that he was too orthodox for a unit such as the Mobile Division. Hore-Belisha, however, now weary of the controversy and wary of overruling the CIGS for the second time, was inclined to accept the offer.[42] He rebuffed Liddell Hart's further efforts to give the command to one of the tank advocates, and, on November 3, the changes were officially made. Brooke was given command of the Mobile Division, and Hobart took his place as DMT. At the same time, Pile was given command of the antiaircraft division then being formed, a post that, while extremely important, was hardly befitting his potential and experience with mechanized units.[43]

The outcome of this struggle was hardly a victory for the tank advocates.[44] True, Blakiston-Houston's nomination to command the Mobile Division had been blocked, but his replacement by Brooke, an artillery officer with no prior experience with armored fighting vehicles, was scarcely an omen that the Mobile Division would be trained along the lines they desired. The power of the army to block an appointment despite the fact that it had the support of the secretary of state for war was established. The struggle also demonstrated that the weakest chink in the tank advocates' armor was Hore-Belisha. Liddell hart's day-by-day account of the struggle reveals that Hore-Belisha had neither an adequate understanding of the radical nature of the theory of armored war nor the realization that half-measures were little better than no mesures at all. He was all too willing to accept the nomination of Brooke to the post as a compromise candidate without considering Brooke's complete lack of experience with mechanized units. For a man who felt a total lack of knowledge of military affairs was not a serious handicap in holding the position of secretary of state for war, it was not, perhaps, a surprising attitude.

The drawn-out affair over the Mobile Division made it more apparent than ever that effective cooperation between Hore-Belisha and Deverell was rapidly becoming impossible. On October 1 Hore-Belisha brought up the possibility of replacing Deverell, and, as the month and the controversy wore on, he became more convinced that the CIGS would have to go. To replace only Deverell, however, was not sufficient: the army was run by the Army Council and to bring about any effective change, the composition of the whole body would have to be altered. As a result, Hore-Belisha and Liddell Hart concluded that, in addition to replacing Deverell, the AG,

General Sir Harry Knox, would also have to go, the post of MGO would be replaced by a director general of munitions production (who was to be a civilian), and the post of deputy CIGS would be revived.[45] Thus, of the four military members of the Army Council, two would be replaced outright, and a third would have his post eliminated.

On November 23, Hore-Belisha informed Chamberlain of his plans and received the prime minister's approval two days later. On November 30, he received the support of the king and the following day sent letters to Deverell and Knox requesting their resignation. In accordance with earlier discussions between "the Partners," these requests were based on the generals' opposition to the policy of "limited liability" rather than their intransigence over the question of appointments and modernization, although in their own minds the latter were the true reasons. This, they felt, would remove any possible stigma of favoritism to the changes by placing them on the level of policy. To the public, however, the retirements were represented as a means of facilitating the promotion of younger men.[46]

The sudden announcement of these dramatic changes came as a shock to both the army and the public. General Pile recalled the "whole army was at six's and seven's."[47] Equally surprising were the new men who came in. Deverell was replaced by General Gort, Hore-Belisha's military secretary, and Knox replaced by Major General C. G. Liddell. Both men were considerably younger than their predecessors—at fifty-one, Gort was thirteen years younger than Deverell—and the newly created post of deputy CIGS was given to Major General Sir Ronald Adam, who was only fifty-two. In all, the average age of the military members of the Army Council was reduced by a decade.[48]

The changes wrought by the army purge of 1937 were undoubtedly necessary and beneficial to the army in many ways, but they did little to bring the needed tank experts into key positions in the War Office. Gort, an infantryman who had entered the army in 1905 from Sandhurst, was considered to be an outstanding leader and generally progressive, but he had had no experience whatsoever with tanks. The same was true of Adam and Liddell. Martel, as assistant director of mechanization, and Hobart, as director of military training, would be of great assistance in furthering the cause, but they could not compensate for any serious deficiencies further up the ladder. Liddell Hart had brought these weaknesses to Hore-Belisha's attention and had also suggested Pile as CIGS, but the secretary was apparently afraid the appointment of one of the tank advocates would be too drastic a move and would cause too much of an outcry in the army. Gort, he felt, was the best all-around choice; he was a young, aggressive, and, according to the then-prevaling view, progressive officer from one of traditional arms. He could thus be expected to retain the confidence of the army as a whole while pushing through reform.[49]

Liddell Hart continued to press for the better use of the tank advocates but with rapidly diminishing success. He raised the point immediately after Gort became CIGS and again at the end of December, when he reported that he was continually being asked if people with tank experience were being brought into key positions in the army. Regretfully, he noted, he was having to answer in the negative. When Hore-Belisha replied that he felt had gotten all of the tank advocates "good jobs," Liddell Hart strenuously objected. Martel and Hobart, it was true, were placed where they could do good, but Lindsay remained in obscurity in India, Broad was major general in charge of administration at Aldershot, and Pile was in command of the 1st Anti-Aircraft Division. This misuse of talent, he contended, could only by the result of continued prejudice against the tank advocates and was extremely harmful.[50] Hore-Belisha, however, refused to be moved. In part, he was apparently growing resentful of his dependence on Liddell Hart and the latter's insistence on so many issues, but his refusal was also the result of the belief that, having changed the Army Council once, he now had to accept its recommendations. As Liddell Hart recorded their conversation on the subject:

> H.-B. felt that his attitude to the new regime must be different than what he had taken towards the old. The latter had deceived so palpably and lost his confidence. Having chosen the men to put in their places, he must show his confidence in them—just as he had let Miss Sloane (his personal secretary) choose her assistant, Miss Fox. The last Selection Board had given him a good case by their ridiculous choices—such as Blakiston-Houston. . . . He said he had exceeded his province as S. of S. [secretary of state for war] under the last regime and wanted to avoid doing it with the new. . . . He said some compromises were inevitable—I said not flagrant injustices.[51]

This, in effect, ended Liddell Hart's attempts to influence the development of the armored forces by securing key positions for the tank advocates. "The Partners" continued to work together, however, and to bring about many necessary changes in other areas. They accomplished much in the field of troop welfare, where the length of foreign service was reduced and improvements were made in soldiers' living conditions, both of which were essential in attracting recruits. Other steps were taken to improve the quality of the officer corps; Woolwich and Sandhurst were amalgamated, and to reduce the age of senior officers, age levels for promotion were reduced and half-pay was abolished.[52]

Still, none of this directly affected the course of mechanization, and the chances of bringing about the needed changes were increasingly being reduced. In part, this was the result of the conservatism of the Army council, but it was also because of a growing rift in "the Partnership" itself. Hore-Belisha was becoming extremely sensitive about his dependence on Liddell Hart, and in January the latter noted that he "seemed rather sensitive" over a comment by Churchill that he was doing nothing but carrying out Liddell Hart's ideas. Several months later, Hore-Belisha went so far as to imply that he did not even know Liddell Hart! Hore-Belisha was also becoming more irritated at Liddell Hart's intransigence on many issues. In January, he complained to Liddell Hart that he was "damned annoyed" with him and said, "I thought you were rough-handling me" over the question of command of one of the divisions. The issue soon passed, but Liddell Hart averred that he was neither content with partial success nor willing to compromise his goals.[53] Still, Hore-Belisha began to give Liddell Hart's advice less weight and sought him out less often. Their contacts, which had been daily during the autumn of 1937, declined sharply in the spring of 1938, and Liddell Hart began to doubt the wisdom of continuing "the Partnership."

On June 1, Liddell Hart gave Hore-Belisha a note in which he attempted to resolve their differences by airing them in the open. He emphasized that, whereas Hore-Belisha had previously reacted favorably to his suggestions, this was no longer the case. As a result, Liddell Hart expressed his growing conviction that the continuation of the "the Partnership" was no longer wise; it cost him his position as an independent critic, but was not yielding what he considered to be sufficient results in terms of concrete progress. To this, Hore-Belisha had no reply, nor did he contact Liddell Hart again until three weeks later. They spoke again briefly early in July, at which time Liddell Hart again stressed that he was of no value to the secretary unless he was completely honest, but again he received no reply. On July 2, Liddell Hart tried to contact Hore-Belisha, but when he could not be reached, asked for the call to be returned. It never was.[54]

The breakup of "the Partnership" ended the tank advocates' best opportunity to control the development of Great Britain's armored forces. Its failure was the result in part of Liddell Hart's support for the policy of "limited liability," which exacerbated the army's opposition to anything associated with his name, but the fundamental reason was the army's continued adherence to the strategy of attrition and the cavalry concept of tank warfare. Liddell Hart was doubtlessly correct in his charge that the War Office consciously avoided placing the tank advocates in key positions, but this was not different than what he was trying to do at the same time. There was nothing sinister about this; it was simply a case of two groups, each convinced of the correctness of its cause and seeking to control the future through the appointment of personnel. It is clear that throughout this phase of the struggle Hore-Belisha was a major stumbling block to the tank advocates' hopes. His failure to appreciate the necessity of placing the development of the armored forces in the hands of the tank advocates reflected his lack of understanding of the radical nature of the theory of armored war and positioned him far closer to his military advisors in the War Office than to his "partner." When his growing sensitivity to his dependence on Liddell Hart reduced the latter's effectiveness, their breakup was inevitable. By the spring of 1938, "the Partnership" had largely lost its effectiveness; the break in July was merely a formality.

Final Preparations

The twenty-two months between July 1938 and the outbreak of the battle of France form a single and clearly defined period. Liddell Hart's position was again restricted to that of an independent critic, and the voice of the tank advocates was largely muted within the walls of the War Office. The tenor of the army was set by General Lord Gort and, when he left to take command of the BEF in France, General Sir Edmund Ironside. Neither represented a significant advance over earlier views on the role of armored forces. The strategy of attrition and the cavalry concept of tank warfare, which had emerged in the mid-1930s, continued to dominate the army's thinking right up to the battle of France.

Although Gort had exhibited a progressive attitude toward the armored forces and the need for new thinking concerning their use prior to his appointment as CIGS,[55] his attitude changed sharply soon after he took up his new post. Part of this change was probably no more than a change of appearance—Liddell Hart felt Gort said what others wanted to hear him say, so he spoke in progressive terms when speaking to Hore-Belisha and himself—but part was also the result of the pressure of more conservative officers. Not long after Gort became CIGS, he was approached by many senior and retired officers and told that he must move slowly. Indeed, in a conversation with Liddell Hart, Gort admitted that progress could be quicker, but "he wanted to avoid criticism from more conservative soldiers, especially the retired generals."[56] As a result, progress continued at a pace only slightly more rapid than it had under the previous Army Council. Mechanization of the army as a whole continued, and in October a War Office memorandum declared that the reequipment of the cavalry with light tanks and armored cars could be regarded as completed by April 1, 1939. But no fundamental changes in the army's structure were made or even contemplated. For the armored forces, this simply meant more of the same. In addition, Gort also set about the task of shutting out Liddell Hart's influence as much as possible, and the one outspoken tank advocate in the War Office, Hobart, was made to feel very unwelcome and was shortly replaced.[57] If nothing else, these events proved the wisdom of Liddell Hart's insistence on putting tank policy solely in the hands of the leading tank advocates for they were the only ones who could correct the situation. It was an empty victory.

To many, however, these problems were obscured by the formation of the Mobile Division in the spring of 1938. As a result of further study and discussion, the army had decided to transfer the motorized infantry battalions from the cavalry brigades to divisional support troops and replace them with additional light-tank regiments and to eliminate one of the mixed battalions of the Tank Brigade. Thus reorganized, the division now consisted of two mechanized cavalry brigades composed of three regiments of light tanks each, the tank brigade, composed of a light battalion and two mixed battalions and divisional troops of two mechanized Royal Horse Artillery regiments, two motorized infantry battalions, and a Field Squadron, Royal Engineers.[58] All of these changes added to the flexibility of the division as a whole and were clearly beneficial, but the formation remained ill-balanced and unmanageable. The proportion of seven light-tank battalions to only two mixed battalions did not give it enough punch; the proportion of nine tank battalions to only two infantry battalions made the division too tank-heavy; and the total number of tanks—over 600—made it almost completely unmanageable no matter how it was organized. Clearly, much remained to be done.

Meanwhile, the army was still trying to overcome the problems of tank design and production. By mid-1938, it had made considerable progress in several areas: eight new Mark VI light tanks a week were being produced, the design for the new Mark II infantry tank had been accepted and was about to go into production, and the medium class of tanks had been abolished and replaced by the cruiser tanks. The army continued to have considerable problems with the latter, however. Because of their requirements, the light tank and infantry tank presented relatively few technical problems; the former was a light, fast machine that required little armor and the latter was a slow, heavily armored machine that required only limited speed. The cruiser tank, however, had to combine both of these requirements and was thus more difficult to design and to produce.[59] The first cruiser, the Mark I (A.9) had been designed in 1934, but it had been plagued by technical problems and progress had been extremely slow. At the same time, a more heavily armored tank along the same line was being developed as an infantry tank, but improvements in antitank weapons made it too light for this role, and it was redesignated as the Mark II cruiser tank (A.10). Then, in the fall of 1936, a British military mission to the Russian Army maneuvers observed a new tank based on a suspension system developed by an American, J. Walter Christie. Greatly impressed, they returned home and urged that the system be incorporated into a new cruiser. In July 1937, the War Office decided to go ahead with the new tank, designated the Mark III cruiser tank (A.13), but because there was no possibility of going into production with the new model until 1939, development of the Mark I and Mark II models continued as stopgaps since they were much closer to delivery. Thus, by the summer of 1938, the British Army was in the process of developing three versions of cruiser tanks in addition to the light and infantry models, an obvious strain on its limited resources that it could ill afford at this time.[60] In part, these prob-

THE MOBILE DIVISION
1938

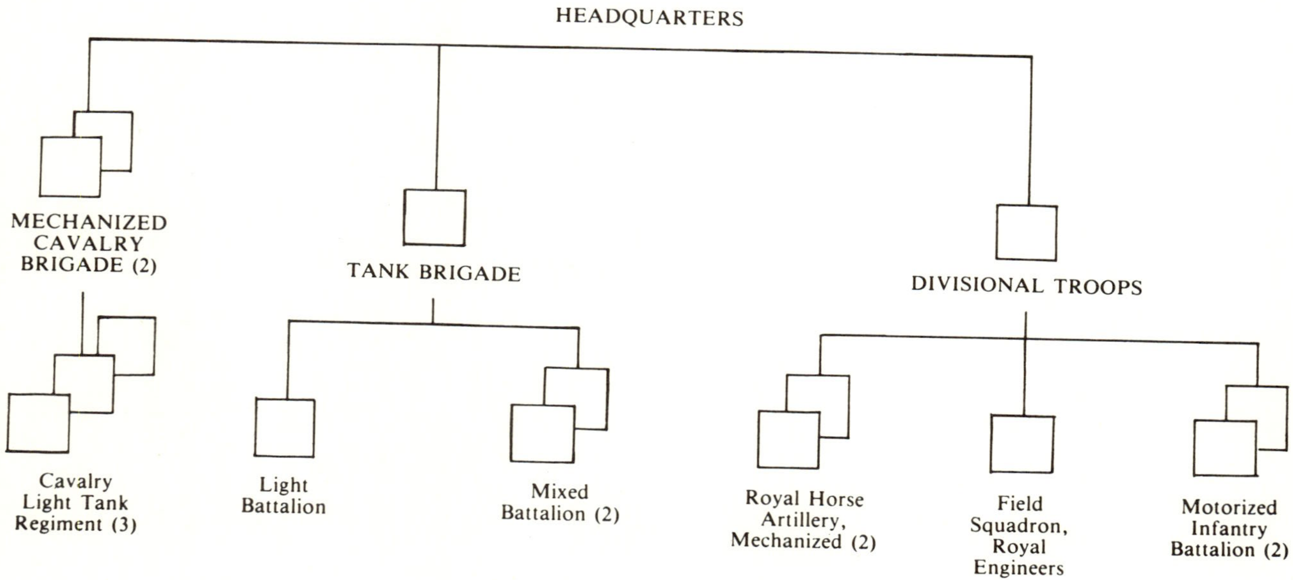

lems were the result of a lack of foresight on the part of the army. The Christie suspension system, for example, had been developed in the late 1920s, but had escaped the notice of even the tank advocates until 1936. Once the decision had been made to go ahead with the Mark III cruiser, continuation of the work on both earlier models was questionable; in fact, because of Martel's efforts, the development of the Mark III proceeded so rapidly that it began to reach units before either of the earlier models. More importantly, however, these problems were another baneful consequence of the army's continued adherence to the cavalry concept of tank warfare. The War Office files in which these decisions are recorded reveal remarkably little confusion regarding the army's tank requirements; rather, they reveal that the army had a very clear perception of the different roles they wanted tanks to play and how these roles required tanks of different design. Like the organizational weaknesses of the Mobile Division, the multiplicity of tank designs can be directly attributed to the army's continued perception of tanks as specialized vehicles performing a variety of functions—just as horsed cavalry had done in the past.

In the meantime, the army had received approval to mechanize two more cavalry regiments[61] and had further revised the establishments of the existing mechanized cavalry regiments and tank battalions. On May 2, 1938, the War Office forwarded a new tank program reflecting these changes to the Treasury Inter-Service Committee. It called for the mechanization of twelve cavalry regiments with light tanks, the replacement of the light and medium tanks in the Tank Brigade with light and heavy cruiser tanks, the addition of a cruiser battalion in Egypt to be taken from one of the army-tank battalions in the expeditionary force, the establishment of three army-tank battalions equipped with infantry tanks in that force, and the expansion of the Royal Tank Corps to eight battalions to enable it to provide the organization for both independent operations and direct support. The total number of tanks called for in the new program, including war reserves, was 2,256, of which 1,331 were light tanks, 551 were cruiser tanks, and 340 infantry tanks. Although this proposal represented a reduction of 501 tanks at a savings of 7 million pounds from the previous request, this change was met almost totally by reducing the number of infantry tanks and thus represented no change in the strength of the independent tank forces.[62] On May 4, the Treasury Inter-Service Committee approved this program and authorized the War Office to place orders for the balance of tank requirements that had not previously been approved.[63]

The 1938 training season obviously did not reflect any of these changes. The previous December, the War Office had decided that, because of the reorganization and reequipment of the army, a two-year training cycle would be established, the first being devoted to the techniques of controlling units and the second to the higher direction of war. As 1938 began the cycle, it was restricted to brigade and subordinate-unit training.[64] For senior officers, a series of map exercises were held, but these afforded little practical experience, serving mainly to confirm that the army had still not learned how to direct armored forces. In May, Liddell Hart told Hore-Belisha that the Mobile Division was making "moves on paper that could hardly be made in war," and later that month Hobart noted, "The Mobile Division is still trying methods that the Tank Brigade had tried and discarded as unworkable three or four years ago."[65]

The tank advocates' hopes were dimmed even further when Hobart was transferred out of the War Office in September. Although his new position as commander of the Mobile Division to be formed in Egypt was suited to his talents, he firmly believed that he was being transferred because he had caused too many waves as director of military training. It had become clear to him that he "was not really looked on with favor by the new lot at the War Office for some time,"[66] and he had additional doubts as to whether the army was really serious about forming a new mobile division in Egypt. Rather, it seemed as if they were merely throwing together what units were available without paying much attention to the experiments and training that had gone on in England since 1934.[67] The organization of the new division seems to bear this out, although the army undoubtedly felt it was doing the best it could under the circumstances. As originally constituted, the division was composed of a cavalry brigade of one armored-car regiment and two cavalry light-tank regiments, a tank brigade of two medium-tank battalions, and a regiment of mechanized Royal Horse Artillery.[68] All of these units suffered from obsolete or inadequate amounts of equipment, and the division as a whole lacked any signal, engineer, supply, medical, or ordnance support. In addition, it contained no infantry until a battalion of the King's Royal Rifle Corps was added in the spring of 1939.[69] Although this situation was far from ideal, the importance of the Suez Canal in British strategic thinking made it an assignment of prime importance and, incidentally, one that Hobart filled superbly. From September 1938 until late 1939, he trained the division in the techniques of armored and desert warfare so that when the war broke out, it was, in the words of General Sir Richard O'Connor, "the best trained division I have ever seen."[70] As the Seventh Armoured Division—the famed Desert Rats—it lived up to this evaluation in the campaigns of 1940–41 and won the reputation of being one of the finest units in the British Army.

Valuable as this work proved to be, Hobart was undoubtedly correct in his belief that his transfer was the result less of his leadership ability than it was of his efforts to make as powerful a case as possible for tank warfare in the army training memoranda issued under his direction.[71] Until Hobart had become director of military training, these memoranda had largely ignored the training of tank units except in the narrowest areas such as communications and movement procedures. Beginning in Training

Memorandum No. 20, issued in April 1938, and continuing in No. 21, issued the following December, however, their role and importance were greatly magnified. Tanks were presented not as a subordinate element of the army, but as a major component. Their offensive roles of exploitation, pursuit, and penetration behind enemy lines of up to a hundred miles were covered at length. The guiding principle behind all these operations was the tank's mobility, which would enable it to threaten alternate objectives, bypass isolated enemy positions for slower units to mop up, and cut off retreating enemy units rather than simply force them back at a faster pace than they otherwise would have gone.[72] On the defensive, they were depicted as a mobile reserve capable of moving rapidly to close any breach the enemy might make. The memoranda particularly stressed that, in any of these operations, "commanders needed to have an entirely different technique and attitude of mind" in order to be successful.[73] It is obvious there was nothing novel in the substance of these memoranda, but the fact that such ideas were expressed in official War Office publications was a striking departure from official doctrine and undoubtedly disturbing to many. Hobart's transfer left the tank advocates without a voice in the major decision-making areas of the War Office and ended the last significant challenge to the cavalry concept of tank warfare.

In early 1939, the Mobile Division was redesignated the First Armoured Division and underwent another major change in organization. The army had recognized the weaknesses inherent in the 1938 organization for some time, for the division was obviously too large and contained to high a proportion of light tanks, and had wanted to reorganize it into two considerably smaller divisions.[74] But administrative problems, particularly in housing, had prevented any changes until now. As reorganized, the division now consisted of a light armored brigade of three mechanized cavalry regiments equipped with light tanks, a heavy armored brigade (the old Tank Brigade) composed of three mixed light- and heavy-cruiser battalions, a motorized-infantry battalion, and artillery and engineer support. Its strength was set at 213 cruiser tanks and 108 light tanks, solving both the problems of size and too high a proportion of light tanks.[75] Unfortunately, the elimination of one of the infantry battalions reduced the tank-infantry ratio from nine to two, which was already too tank-heavy, to six to one, reflecting both the army's continued belief in the cavalry concept of tank warfare and the tank advocates' belief that this was the only means of maintaining the division's independent role. Although these changes were on the whole beneficial, there is no little irony in the fact that this point—virtually the only one on which the army and the tank advocates could agree—proved to a major weakness of British armor in the early stages of World War II.

These changes in organization did little to improve the status of the division, however, for it remained woefully lacking in equipment. Light

THE ARMOURED DIVISION
1939

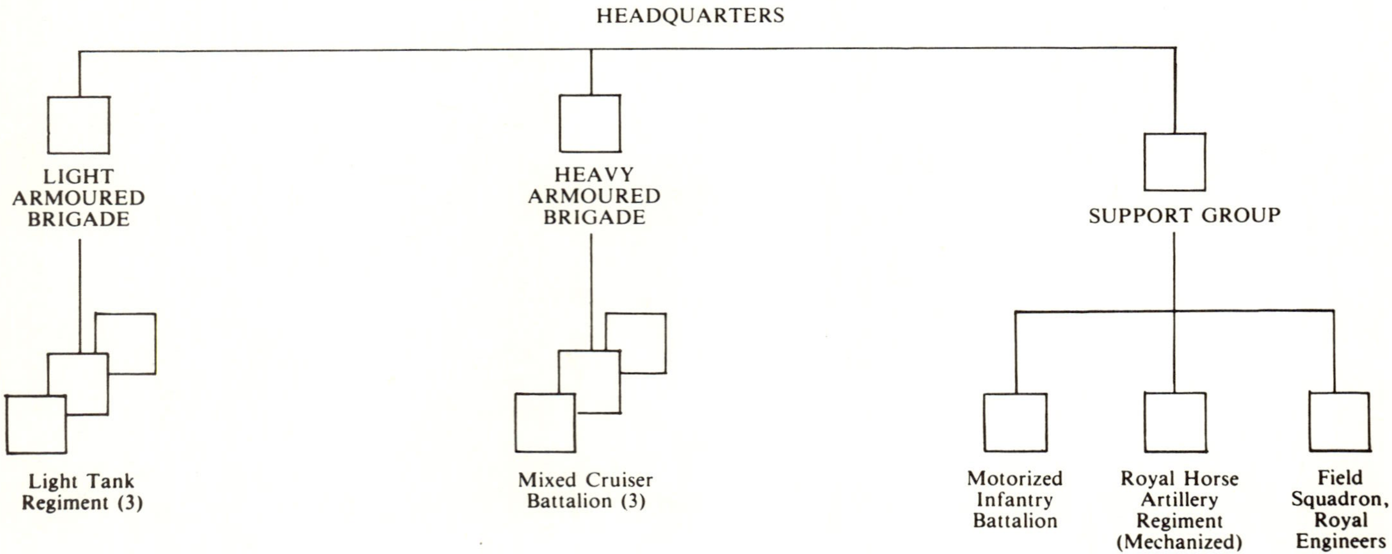

tanks were now reaching the division in some numbers, but the first cruisers had yet to come off the assembly line, which meant that the heavy brigade had no modern equipment whatsoever.[76] Moreover, Brooke was replaced at this time by Major General Roger Evans, a cavalryman who not only had no prior experience with mechanized forces but also, according to the *Army List* for January 1939, was a staff officer who had never held a command before.[77] As Pile noted a few months later, he was "very much out of his depth with the Mobile Division. He apparently recognizes this himself, favoring the appointment of a Mobile Corps Commander so that he may have someone above to provide guidance."[78]

In the meantime, the army was still struggling with the government's policy of "limited liability." On February 20, it achieved a major success when the government approved a comprehensive plan to equip fully both the regular and the territorial forces on a scale for Continental service. The plan was a product of an army proposal submitted the previous October that had been heatedly debated both in the War Office and the government committees to which it was forwarded. Originally, the army had justified its request on the grounds that Great Britain should be prepared to send a force to the Continent in the event of war, but this had aroused so much opposition that Hore-Belisha finally insisted that the proposal be sent forward with no justification.[79] In this form, he argued strenuously for its adoption, winning praise from General Pownall, scarcely one of his greatest admirers, who confided to his diary: "Let us give the devil his due. H.-B. is now fighting gallantly for the field force, indeed for the army as a whole. He is fresh and active and taking much more helpful interest in things."[80] As approved, this program established an expeditionary force of two armored and two infantry divisions capable of immediate deployment overseas, two more divisions for deployment in two months, and four territorial divisions for deployment in six months. In addition, the nine remaining territorial divisions would receive a full complement of equipment for training.[81] In all, this would raise the army to a strength of nineteen divisions, of which three would be armored, for a ratio of just over one to five.

The program did not remain in effect for long. In March 1939, Hitler occupied all of Czechoslovakia, causing the French and British to reevaluate their positions, and on March 29, Chamberlain announced the government's decision to bring the Territorial Army up to its war strength of thirteen divisions and then to double each division, giving the army a total strength of six regular and twenty-six territorial divisions. Of this number, four would be armored divisions (three in the expeditionary force and one in Egypt) and twenty-eight would be infantry divisions, for a ratio of one to seven.

On April 4, the War Office announced the formation of the Royal Armoured Corps, to be composed of the eighteen regiments of mechanized cavalry and eight battalions of the Royal Tank Corps, henceforth designated the Royal Tank Regiment.[82] This decision was the product of a long study by a War Office committee appointed in the spring of 1938 to study the future organization of these two arms. As the committee frankly admitted in its report, the reason for this action was that the cavalry was in trouble. The continued existence of so many horsed regiments until the mid-1930s had given it the appearance of an archaic branch of the army with an uncertain future and had made recruiting extremely difficult. Moreover, those regiments which had become mechanized found the transformation particularly difficult because they lacked the personnel and facilities to train their people in the use of the new equipment. The Tank Corps, on the other hand, was particularly well situated in both of these areas. Its training program was well established and centralized, and because it was perceived in the popular mind as one of the most progressive and dynamic branches of the army, it actually had a surplus of recruits. This had enabled it to raise its physical and educational requirements to the point where it was accepting only the best of the applicants who came forward.[83] The tank advocates' solution to this problem was, naturally, that the army should abolish the cavalry and expand the Royal Tank Corps, but this course was not even considered by the committee. Instead, it opted for the creation of an "umbrella" unit under which both branches could coexist in a sort of separate but equal status. The tank and cavalry regiments would each maintain their own officer lists to preserve the social distinctiveness of the latter, but enlisted recruits would enter the corps as a whole and receive their basic training at a centralized school. They would be assigned to regiments only after completion of this training, and they could be transferred from one unit to another if the need arose.[84] This arrangement undoubtedly improved the efficiency of the mechanized cavalry and followed logically from the army's operational doctrines and regimental traditions, but also served as a further check to broader acceptance of the theory of armored warfare and the employment of the tank advocates in key positions controlling the expansion of the army's armored forces.

Three weeks later, the situation facing the army was radically transformed when the Chamberlain government announced the first peacetime conscription in British history. The immediate reason for this decision was the need to secure additional personnel to man the antiaircraft defenses, but the pressure for such a move as a demonstration of Great Britain's commitment to its allies had been growing for some time.[85] Whatever the validity of these considerations, its immediate effect on the army was anything but positive. Liddell Hart naturally opposed this move, seeing it as another demonstration of the manpower obsession that had dominated military leaders since the time of Napoleon. Its effect, he later recalled, was that "Britain's military and industrial effort was now geared to the production of a mass army of infantry instead of to a high quality mechanized force."[86] This argument was continuation of his earlier criticisms of the War

Office plans to modernize the entire regular army rather than concentrate on the Tank Corps. As noted, these earlier charges were questionable because of the small size of the regular army, but this was not the case now. Conscription vastly increased the strains on the British economy and forced those tank programs which had already been approved to take a back seat to the production of equipment for the influx of new troops. As the Official History of the war noted, "clothing and other personal equipment, unlike guns and tanks, had to be available at the very outset of mobilization, for soldiers had to be clad, fed and housed even before they could be trained in the use of weapons."[87] In sum, while there was a definite need for Great Britain to make a stronger expression of its determination to stand up to further Nazi expansion, the wisdom of the particular response it adopted is highly questionable.

One of the most immediate consequences of this decision was to complete the disruption of the army's two-year training cycle established in December 1937. According to that program, the 1939 training season should have been devoted to the higher direction of war and would have involved maneuvers on the scale of 1925 and 1935. Whether it would have been possible to hold these exercises had conscription not been declared is uncertain because of continued shortages of equipment, but this decision made them impossible. As a result, the only training undertaken for senior officers was a series of command map exercises that, in Liddell Hart's opinion, served one only to reveal "a lack of grasp of the problems of mechanized movement on the part of the potential commanders of the Field Force."[88] The effect of this lack of training cannot be overemphasized. Since 1918, the British Army had conducted large-scale maneuvers only in the two years just noted, and none of the senior officers in 1939 had played a major role in either of them. Consequently, the army had no one with any practical experience in controlling and coordinating the movement of large numbers of troops in the field. Moreover, whatever experience may have been gained by officers serving in lesser postions in the 1925 and 1935 maneuvers would have been largely obsolete by 1939, because the army's mobility had been radically transformed by mechanization. This situation was particularly striking with regard to the armored forces, as the army had never fielded anything larger than the Tank Brigade. In short, Great Britain was preparing to go to war with an army whose equipment and senior officers were completely untested in the highest level of operations even by the mild standards of peacetime maneuvers. That no one was concerned about this was a further indication that the army still failed to see the revolutionary potential of the changes they had wrought.

The problem was compounded by the fact that, when the war broke out in September 1939, Great Britain's tank forces still existed largely on paper. The army had seven mechanized cavalry regiments equipped with light tanks and one equipped with armored cars, but, outside this, it had no units fully equipped with up-to-date tanks. The tank battalions of the First Armoured Division's Heavy Armoured Brigade had only six cruisers each, and, while one army-tank battalion for direct support was sent to France, it was equipped with Mark I infantry tanks, which were so slow and mechanically unreliable that General Broad could not use them in the Aldershot tattoos that summer. The army had only two Mark II infantry tanks ready for service.[89] Thus, despite the plans and preparations of the three previous years that had called for the production of over 2,000 tanks, the British Army could muster only 196 light tanks, 50 obsolete infantry tanks, and 38 armored cars in operational units when the war broke out.[90]

Yet this situation did not lead the army to radically change its tank-production program. As laid down in a series of military conversations the previous spring, Anglo-French strategy was based on a long period of static warfare during which the Allies would gradually build up their strength to the point where they could then go on the offensive. As a result, the pace of war production increased only slightly. In the words of the *Official History,* it was a change from "disturbed peace" to "dormant war."[91] Within the War Office, a similar atmosphere prevailed. The Treasury continued to impose strict peacetime budgetary proceedures and considerations, and delays quickly developed in the production of all types of weapons and supplies.[92] The only dramatic change produced by the war was, in fact, the assignment of Lord Gort to command the BEF and his replacement as CIGS by General Sir Edmund Ironside. These appointments came as a shock not only to the army, but also to the men involved; Ironside had been told he would command the BEF in the event of war and expressed serious reservations concerning his new appointment as he had never served in the War Office before.[93] He accepted it, however, because he felt it his duty to go where he was called.[94] This change did not have any significant effect on the armored forces; although Liddell Hart had written only two years earlier that Ironside was the only senior army officer who had made an in-depth study of mechanization, he had since noted the general's decline as he grew older. Moreover, as the 1938 Command Exercises had demonstrated, Ironside's grasp of the use of the Armoured Division was extremely weak.[95] Curiously, though, Ironside tried to have Fuller brought back to active service to serve as his deputy, only to have his request refused by Chamberlain because of Fuller's recent fascist activities.[96]

During the fall, Great Britain began laying plans for an even greater increase in the size of the army and, notably, Hore-Belisha was among those who led the way in this movement. On September 8, the Land Forces Committee, a special group designated by the War Office to study the question of the size of the army, recommended the formation of a fifty-five–division army by the end of the second year of the war. The following day, the Ministry of Supply reported that such an effort would incur many problems the country might not be able to overcome. When he received the report, Hore-Belisha, very much annoyed, wrote at the top: "Why we should not have an Army! by the Ministry of Supply." Later that day, he recorded in his diary:

> I decided to put in a Memorandum on the size of the Army at once to the Cabinet, asking for the objective of 100 divisions. My purpose was to set out what the Army objective should be on the assumption that we are now engaged in a war of total effort. Our aim should not be less than to prepare production capacity for 100 divisions.[97]

Hore-Belisha stressed this point again ten days later, when, upon receiving a memorandum from the chancellor of the Exchequer saying that anything more than a twenty-division army "would reduce the country to bankruptcy," he wrote to Gort that they must fight this idea.[98] Given the position in which Great Britain found itself at this time and the example of the Polish campaign, it was inexcusable that such emphasis continued to be placed on numbers of divisions to the neglect of tank forces. It was simply another indication that he never understood the theory of armored war.

Even with this emphasis on increasing the total size of the army, however, Great Britain was able to increase its tank production substantially following the outbreak of the war. In fact, for the last four months of 1939 and first four months of 1940, British tank production actually surpassed that of the Germans.[99] Unfortunately, too high a proportion of this number continued to be light tanks. By the spring of 1940, only one more army-tank battalion had been dispatched to France, although this one was equipped with Mark II infantry tanks, which were a marked improvement over their predecessor. As for the First Armoured Divison, it was reorganized in March to include an additional infantry battalion and eliminate the distinction between the light- and heavy-tank brigades by equipping both completely with cruiser tanks.[100] Both of these changes were extemely beneficial from the perspective of the theory of armored war, but they delayed the deployment of the division even longer. It was finally dispatched to France only after the German assault began, and even then it was not fully equipped. Of its 284 tanks, only 150 were cruisers and many of them lacked such equipment as radios and armor-piercing ammunition.[101]

Despite the absence of the armored division, however, the British high command remained optimistic regarding its ability to withstand any German onslaught, although they readily admitted it was not prepared to go on the offense.[102] By the beginning of May, the strength of the BEF stood at nine infantry divisions, two light armored reconnaissance brigades of two divisional cavalry regiments each, one army tank brigade of two army-tank battalions, and four separate divisional cavalry regiments.[103] Its tank strength stood at almost 350 machines, with the expectation that this would nearly double before summer with the addition of the First Armoured Division. Even more significant from the army's point of view, it was the only army in the field that was wholly mechanized; not a single horse was to be found anywhere. General Pownall, then chief of staff of the BEF, reflected their optimism when he confided to his diary:

> I know of no part of Europe, bar the mountains, so well adapted to dealing with anti-tank defense, we have a continuous A. T. [antitank] obstacle along our front, another behind us in the La Bassei-Donai canal, we shall construct one between these two, and if we advance to the Escaut . . . [in Belgium] we have yet another, with the above named three behind us. What more can the heart of man desire? All that apart from the fact that we have a higher proportion of A. T. weapons (even excluding the A. T. rifles) than either German or French divisions.[104]

The British Army and the German Blitzkrieg

Undoubtedly one of the major reasons for this optimism was the widespread conviction throughout the British Army that they were at least as advanced as any other army and ahead of most in the use of armor. This conviction was based on a broad study of foreign military publications. Beginning in January 1938, the *Royal Tank Corps Journal* published a regular series of articles entitled at first "The Foreign Press" and later "Foreign Views and News," in which it examined the organization, tactics, and training of foreign tank forces. This was supplemented by specific studies and reviews of foreign books when deemed appropriate and by a scattering a similar articles in other journals.

While it is difficult to generalize about the opinions presented in these articles, as they were often conflicting, two themes are apparent. The first is that there was a widespread belief that armored forces were still very much in the testing stage. It was obvious to all that both tanks and antitank guns had undergone considerable technological changes since the end of the last war, and that without any practical battle experience on a large scale it was impossible to predict what they might or might not be able to do in the future. The experience of the Spanish Civil War was limited, in the *Royal Tank Corps Journal*'s view, by the fact that tanks were employed only in small numbers and under peculiar conditions and that they were often misused.[105] Nonetheless, there were some lessons learned, although only on a tactical level. In particular, the journal stressed that the war demonstrated (1) the need for the closest cooperation between infantry and the tanks assigned to support them; (2) the impossibility of "extemporized" forces carrying out independent tank operations; and (3) the use of surprise and tanks in mass as "prime factors in victory in the case of all tank operations, whether carried out in cooperation or independently."[106] Beyond such obvious generalizations, however, the journal would not go.

This lack of certainty was even more apparent in the journal's coverage of the controversy between the advocates of highly mobile tank warfare and those who wanted to restrict tanks to the pace of infantry. Considerable attention was given to both sides in this argument, and, while the British recognized that the Germans were generally following the former course and the French the latter, exceptions in both armies were noted.[107] Moreover, no special weight was attached to the German doctrine. The German Army was certainly respected as a formidable force, but it had been slow to develop tanks in World War I and, because of the limitations of the Versailles Treaty, had not entered the tank race until 1935. As a result, while the British did not ignore developments in Germany, they gave no indication that they believed the Germans had anything to teach them. As one article that described the contrasting views of the Germans and French concluded, "The much discussed and much argued question, whether the tank can exist as the main weapon in battle or whether it must remain as an auxillary weapon for the infantry, can only receive a clear answer in war."[108]

The second theme that emerges from this coverage of foreign tank development was the British failure to perceive any strategic implications in the German acceptance of mobile tank warfare. Although the *Tanks Corps Journal* regularly reported on articles published in its German counterpart, such reports dealt exclusively with organization, tactics, and equipment. It gave particular attention to German tactics in the "deliberate attack," that is, assaults on carefully prepared positions, and it explained how the Germans believed such operations could only succeed if they penetrated the entire defensive zone before enemy reserves could be rushed to the scene. The key, they believed, was for waves or echelons of tanks to lead the attack on successive defensive positions, with support being provided by motorized infantry, artillery and antitank units following up as rapidly as possible. In any event, the advance of the tanks should never be held up by waiting for support before moving to the next position. In comparing these tactics to those of the French, the journal stated that "risks, which no Frenchman would entertain, are encouraged in the movements of the Formations, for the tank is considered as their primary offensive weapon."[109] The journal gave particular attention to the writings of an Austrian general named Ludwig Ritter von Eimannsberger, but, curiously, it never once referred to the work then being done by Heinz Guderian, who has since been recognized as the father of the Panzer forces.[110] Even more curious was the fact that the British paid scant attention to what the German armored forces would do once they had broken through an enemy defensive zone. One article describing Eimannsberger's tactics of penetrating such a zone simply concluded that the last phase of the assault would be "the advance of tank units through the gap thus made and their independent operations."[111] Other articles were only slightly more specific and described "independent" tank operations as generally similar to cavalry operations in the eighteenth or early nineteenth centuries.[112] That such operations could be more than this was never considered.

This misperception of the German tank doctrine also found its way into British intelligence estimates. Again, they demonstrate that the British Army was aware of German tactics in penetrating a defensive zone, but woefully unaware of what might follow. In 1938, the British military attaché in Berlin reported to the War Office:

> The Germans believe that the power of penetration and mobility of the older arms is not sufficient to enable them to carry an attack so rapidly and to such a depth into the enemy organization that he will not have time to take countermeasures. This, the Germans claim, can only be achieved by armoured forces.[113]

In October 1939, an estimate of German intentions prepared by the staff of the BEF in France brought out these same points while providing more detail as to how they might be applied in the situation which then existed. It emphasized that the purpose of the German armored forces was "to break through and reach back areas before the enemy's defense measures are put into effect" and that "on no account will the launching of the armored forces be delayed on the account of the holding out of islands of resistence." Unfortunately, the report then described this method of attack as "infiltration" and predicted it would be executed by an assault at approximately twenty places along the entire British front by combat groups with approximately 120 tanks each.[114] What the British had done was to go back to Fuller's Plan 1919; they envisioned a tactical operation in which tanks would replace the infantry combat teams the Germans had infiltrated behind the Allied lines during their 1918 offensives. To counter this, the BEF deployed tank battalions and supporting infantry in reserve to deal with any German armored forces that might get through. Based on the 1934 exercises in which Lindsay's mobile force had been defeated, the concept was sound, providing the Germans did what was expected of them. Unfortunately, they did not.

Much has been written about the German blitzkrieg in the opening campaigns of World War II and, to a lesser degree, the extent to which the Germans were influenced by the writings of Fuller and Liddell Hart. Although the British Army showed no awareness of the fact at the time, many of Fuller and Liddell Hart's writings were translated into German in the 1930s, where they inspired a small but important group of tank advocates in that country. Heinz Guderian admitted after the war that

> it was principally the books and articles of the Englishmen, Fuller, Liddell Hart and Martel, that excited my interest and gave me food for thought. These far-sighted soldiers were even then trying to make of the

> tank something more than just an infantry support weapon. They envisaged it in relationship to the growing motorization of our age, and thus they became the pioneers of a new type of warfare on the largest scale.
>
> I learned from them the concentration of armour, as employed in the battle of Cambrai. Further, it was Liddell Hart who emphasized the use of armoured forces for long-range strokes, operation against the opposing army's communications, and also proposed a type of armoured division combining panzer and panzer-infantry units. Deeply impressed by these ideas I tried to develop them in a sense practicable for our own army. So I owe many suggestions of our further development to Captain Liddell Hart.[115]

There is considerable truth in this statement, as Brian Bond has shown in his study of Liddell Hart's military thought.[116] Certainly in terms of the actual conduct of operations, there were many similarities between the strategy of armored war and the German blitzkrieg: both called for the employment of massed armored formations to achieve strategic paralysis, and both held that this could only be accomplished if such formations were freed from the slow pace of infantry and sent deep into the enemy's rear to disrupt his command and communications network. But beyond this, there were profound theoretical differences between the two doctrines of war.

The German conception rested upon a combination of traditional and revolutionary ideas.[117] Since the Austro-Prussian War of 1866, German military doctrine had been founded on the conviction that modern technology had radically altered the balance betwen offense and defense. Strategically, the growth of railroads enabled commanders to move troops and supplies on a scale and with a speed hitherto impossible and thus greatly favored offense. Tactically, however, technological advances had produced rapid-fire weapons with far greater range than before, which served to greatly strengthen the defense. As a result, the Germans had concluded that future victories would depend upon the successful application of strategic offense and tactical defense; that is to say, victory would go to the army that could advance upon and surround its enemy, leaving him the alternatives of either surrendering or of launching fruitless frontal assaults. This concept was called *Kesselschlacht,* or "cauldron battle."

The *Kesselschlacht* concept worked brilliantly in the Franco-Prussian War of 1870, in which Louis Napoleon was forced to surrender with an entire army at Sedan. In 1914, however, the Germans failed to achieve a similar victory when, after advancing through Belgium and well into France, they were halted on the Marne before the encirclement of the Allied armies was completed. The result was an unexpected and unwanted struggle of attrition and eventual defeat. In postwar studies to determine the cause of their failure, the Germans concluded that the *Kesselschlacht* doctrine was still sound, but that it had failed in 1914 because they had been unable to disrupt the mobilization and redeployment of the French Army, which, in turn, was the result of the slowness of their advance. In other words, for the *Kesselschlacht* doctrine to succeed, a way had to be found to prevent the enemy from retreating before he could be surrounded. It was at this point that the potential of strategic penetration by armored forces struck a responsive chord in the German Army. By sending massive armored formations deep into the enemy's rear, he could be pinned down until the bulk of the German forces completed surrounding him.

Seen in this light, the differences between the German and British conceptions of blitzkrieg war stand out clearly. In the first place, where the British theorists sought to eliminate battles in the traditional sense, the Germans only sought to insure that they would be fought on terms more favorable to them. In the second place, where the British saw little or no role for the traditional combat arms (and thus aroused a great deal of hostility within the army), the Germans continued to see these arms playing a decisive role. They would indeed be the principle element in the final act of the *Kesselschlacht.* And finally—and perhaps most importantly—where the British theorists placed their army in the position of having to admit its conduct of World War I was predicated on unsound principles in order for their ideas to be accepted, the Germans presented no such challenge. In short, the German Army did not have to go nearly so far to accommodate its doctrines to the principles of strategic paralysis implemented by massed armored formations as did the British. Nor did it. By the eve of World War II both armies were preparing to wage war in accordance with principles considered sacrosanct since the nineteenth century, the British according to the strategy of attrition, the Germans according to the strategy of *Kesselschlacht.* This was the key to their employment of armor in the coming campaigns.

Notes

1. Field Marshal Sir Edmund Ironside, *The Ironside Diaries: 1937–1940,* R. MacLeod and D. Kelley, eds. (London: T. and A. Constable, Ltd., 1962), p. 24. When he first heard of the appointment, Ironside expressed doubt regarding Hore-Belisha's ability to bring about the necessary changes, but within a few days, he had changed his mind and wrote, "He will probably be our saving. He is ambitious and will not be lazy like some of the others were."

2. Leslie Hore-Belisha, *The Private Papers of Hore-Belisha,* R. J. Minney, ed. (Garden City, N.Y.: Doubleday, 1961), p. 17. Deverell, after World War II, wrote bitterly of Hore-Belisha, making particular note of the fact that he was Jewish. At one point, he noted Hore-Belisha's "cunningness of a clever Jew." Deverell to Brigadier H. R. Sandilands, August 8, 1945 and January 16, 1947, Sandilands Papers, Center for Military Archives, King's College, University of London. In his private diary, General Pownall made the same point, referring to Hore-Belisha as a "shallow-brained, charlatan, political Jewboy." Sir Henry Pownall, *Chief of Staff: The Diaries* 1:209, entry for 1 May 1939. Brian Bond, who states that he has been assured by several British officers who served in the army during this period that they were unaware of any anti-Semitism, concludes that this was merely "another stick" used to beat Hore-Belisha by those who opposed his policies or methods. In any case, it was bound to affect the secretary's efforts negatively. B. Bond, *British Military Policy Between the Two World Wars* (Oxford: Clarendon Press. 1980), pp. 244 and 347, n.32.

3. Conversation between Hore-Belisha and Liddell Hart, 21 August, 1937, "Note for History," Liddell Hart Papers, A4/16. Liddell Hart's "Notes for History" were a record of conversations he held with close friends and key government officials. They were recorded during the conversation (if on the telephone) or as soon thereafter as possible.

4. Liddell Hart, *Memoirs,* 2 vols., (London: Cassell, 1965), 1 : 5.

5. Ibid., 2 : 72–73

6. "Army Training Memorandum Number 17," dated 17 January 1973.

7. *The Times,* 22 September 1937. Interestingly, Liddell Hart wrote of Ironside in August 1937 that he was the only one of the senior generals who had made a long study of mechanization. "An Evaluation of Senior Army Officers," an unpublished memorandum for Hore-Belisha, dated 28 August 1937, Liddell Hart Papers, A4/16.

8. Hobart to Liddell Hart, 24 July 1937, Liddell Hart Papers.

9. WO Memorandum 57, dated 10 November 1936 and minutes of the Defense Policy and Requirements Subcommittee of 13 November 1936, WO 32/4441.

10. Minutes of the Treasury Inter-Service Committee meeting of November 23, 1936, WO 33/4441.

11. These requests were made in WO Memorandum 70 (which is not in the file but is referred to in WO Memorandum 95) and in WO Memorandum 95. The cost figures were noted in the minutes of the meeting of the Treasury Inter-Service Committee on 4 May 1938, WO 32/4441.

12. Martel to Liddell Hart, 21 August 1937, Liddell Hart Papers.

13. Minutew 48A, dated July 1937, WO 32/4441.

14. Memorandum of 23 July 1937, WO 32/4441.

15. WO Memorandum 145, dated 26 July 1937, WO 32/4441.

16. Ibid.

17. Minutes of the Treasury Inter-Service Committee meeting of 27 July 1937, WO 32/4441.

18. Treasury letter dated 16 October 1937, WO 32/4441.

19. Pile to Liddell Hart, 12 June 1937. Similar remarks are found in letters by Hobart to Liddell Hart on 24 August 1937 and Martel to Liddell Hart 8 August 1937, all in Liddell Hart Papers.

20. Liddell Hart, "The Problem of Mechanization in the Army," an unpublished memorandum dated 23 August 1937, Liddell Hart Papers.

21. Liddell Hart to Hore-Belisha, 4 September 1937, Liddell Hart Papers.

22. Liddell Hart, "Notes for History," 23 August, 4, 7, and 11 September 1937, Liddell Hart Papers, A4/16.

23. Quoted in Liddell Hart, "Memorandum of the Organization and Role of the Mobile Division," a memorandum to Hore-Belisha dated 31 October 1937.

24. "The 'I' Tank from an Infantry Man's Point of View," *Royal Tank Corps Journal* 1 (January 1937): 54.

25. "The 'I' Tank, Another Infantry Man's Point of View," *Royal Tank Corps Journal* 2 (July 1937): 28.

26. "The Employment of a Tank Brigade," *Royal Tank Corps Journal* 4 (July 1938): 4. See also "The Function of a Mobile Division," *Royal Tank Corps Journal* 2 (January 1937): 48–51; "The First Tank Brigade," *Royal Tank Corps Journal* 3 (January 1938): 2–4; "Tanks in the Next War," *Royal Tank Corps Journal* 5 (January 1939): 30–33.

27. Hobart, "Organization of Units and Formations in the Regular Field Force Using Armoured Fighting Vehicles," an unpublished memorandum attached to a letter to Liddell Hart dated 22 September 1937, Liddell Hart Papers. Hobart also recognized the need for infantry and other supporting units to assist tanks in the accomplishment of their mission.

28. Hobart to Liddell Hart, 21 October 1937. As a result of this rebutal, Hobart then submitted a water-down version of this paper which was apparently accepted. Because of the revisions, which considerably compromised the tank advocates' ideas, however, it scarcely represented a victory on their behalf.

29. Liddell Hart, "Organization and Role of the Mobile Division," an unpublished memorandum dated 31 October 1937, Liddell Hart Papers.

30. Hore-Belisha, *Private Papers of Hore-Belisha,* pp. 51–52.

31. Conversations between Hore-Belisha and Liddell Hart on 28 August and 10 September 1937, "Notes for History," Liddell Hart Papers.

32. These matters are covered in WO 32/4195, "Proceedings of the Informal Army Council Meetings from July 1937 to February, 1938."

33. Pownall, *Chief of Staff: The Diaries,* p. 99, entry for 27 January 1936. Italics in original.

34. Ibid., p. 123, entry for 1 January 1938. Hankey told Pownall to deemphasize his opposition to "limited liability" prior to the latter's interview with Hore-Belisha regarding his appointment as director of military operations and intelligence.

35. Ibid., p. 125, entry for 3 June 1938, and p. 190, entry for 20 February 1939.

36. Conversations between Hore-Belisha and Liddell Hart, 10 and 11 September 1937, "Notes for History," Liddell Hart Papers. Deverell's feelings were later expressed to Brigadier Sandilands in letters of 8 August 1945 and 26 January 1947, Sandilands Papers.

37. Conversation between Liddell Hart and Hore-Belisha, 12 September 1937, "Notes for History," Liddell Hart Papers.

38. Conversation between Liddell Hart and Hore-Belisha, 5 October 1937, "Notes for History," Liddell Hart Papers.

39. Conversation between Liddell Hart and Hore-Belisha, 5 October 1937, "Notes for History," Liddell Hart Papers.

40. Liddell Hart to Hore-Belisha, 12 September 1937, Liddell Hart Papers.

41. Conversation between Liddell Hart and Hore-Belisha, 19 October 1937, "Notes for History," Liddell Hart Papers.

42. Ibid., 31 October 1937. He later would become Lord Alanbrook.

43. Ibid., 3 November 1933.

44. Liddell Hart to Hore-Belisha, 12 November 1937, Liddell Hart Papers. He said, "Am not happy about the justice or wisdom of the course being taken."

45. Conversations between Liddell Hart and Hore-Belisha, 1–31 October 1937, "Notes for History," Liddell Hart Papers, A4/16.

46. Ibid., 19 October 1937. The text of the letter sent to Deverell is found in Hore-Belisha, *Private Papers of Hore-Belisha,* pp. 70–71. The letter to Knox was identical.

47. General Sir Frederick Pile, personal interview with the author at the Cavalry Club, London, on 6 August 1971.

48. Hore-Belisha, *Private Papers of Hore-Belisha,* p. 75.

49. Liddell Hart, "Evaluation of Senior Army Officers," an unpublished memorandum dated 8 August 1937; conversation between Liddell Hart and Hore-Belisha, 3 December 1937, "Notes for History," Liddell Hart Papers, A4/16.

50. Conversation between Liddell Hart and Hore-Belisha, 31 December 1937, 3 January 1938, "Notes for History," Liddell Hart Papers. Liddell Hart maintained this position later in Liddell Hart, *The Tanks,* 1:366, but General Broad has stated that in his own case, at least, this was not the situation. According to him, General Deverell told Lady Broad that he had assigned her husband to Aldershot to give him additional administrative experience and keep him in hand so that he could be assigned to India to direct the mechanization of forces there. Unfortunately, the situation did not develop as Deverell had hoped and the appointment could not be made. General Broad also expressed the firm conviction that whatever restrictions were made regarding appointments of the tank experts, they were done, in the eyes of the War Office, for the good of the army. They were, he said, "gentlemen, and would never have acted otherwise." Lieutenant General Sir Charles Broad, personal interview with the author at his home, Beaminster, Dorset, 10–12 August 1970.

51. Conversation between Liddell Hart and Hore-Belisha, 3 January 1938, "Notes for History," Liddell Hart Papers.

52. WO 33/1502, "Army Role for War," dated 10 February 1938; Liddell Hart, *The Tanks,* 1:398–99.

53. Liddell Hart, "Notes for History," 18 January 1938, Liddell Hart Papers.

54. Liddell Hart, memorandum to Hore-Belisha, 31 May 1938; also conversations between Liddell Hart and Hore-Belisha 3 and 4 July 1938, "Notes for History," Liddell Hart Papers, A4/16.

55. Gort to Liddell Hart, 24 October 1937, Liddell Hart Papers, A4/15.

56. Interview with General Broad; conversation between Gort and Liddell Hart, 20 May 1938, "Notes for History," Liddell Hart Papers. Both Broad and Pile later expressed the opinion that Gort was far above his head as CIGS. Broad said, "He really hadn't got a brain," and Pile commented, "He was very much lacking in brains." Interviews with Generals Broad and Pile.

57. Conversation between Liddell Hart and Hore-Belisha, 27 March 1939, "Notes for History," Liddell Hart Papers, A4/16.

58. See chart p. 214. Martel, *Our Armoured Forces,* p. 378. Ogorkiewicz, *Armoured Forces,* pp. 58–59. The cavalry light tank regiments were equivalent to light-tank battalions.

59. Liddell Hart, *The Tanks,* 2: Appendix 5, 486–87. Lists the characteristics of the British tanks of this period from which the following extract illustrates this point:

Type	*Weight*	*Speed*
Light tank MK VI	5.5 tons	35 mph
Cruiser tank MK III	14¾ tons	30 mph
Infantry tank MK II	26½ tons	15 mph

60. Liddell Hart, *The Tanks,* 1:375; Ogorkiewicz, *Our Armoured Forces,* 53–57.

61. W.O. 34/4441, Treasury Inter-Service Committee Minutes dated January 18, 1938.

62. WO 32/4441 Memorandum 343, dated May 2, 1938.

63. W.O. 32/4441, Minutes of Treasury Inter-Service Committee meeting of May 4, 1938, ibid.

64. "Army Training Memorandum No. 19," dated 5 December 1937.

65. Liddell Hart, "Memorandum on the Mobile Division," a paper written for Hore-Belisha, dated 10 May 1938, A4/16, Liddell Hart Papers; conversation between Liddell Hart and Hobart, 24 May 1938, "Notes for History," A4/15, Liddell Hart Papers.

66. Hobart to Liddell Hart, 21 January 1938, Liddell Hart Papers.

67. Hobart to Liddell Hart, 10 July 1938, Liddell Hart Papers.

68. "Army Training Memorandum No. 21," dated 31 December 1938.

69. Liddell Hart, *The Tanks,* 1:402–3.

70. Recently, there have been claims that Hobart merely built on the work of predecessors who had conducted innovative tank training in Egypt for almost a decade. Bond, *British Military Policy,* pp. 182–83; Ronald Lewin, *Man of Armour: A Study of Lieutenant General Vyvyan Pope* (London: Leo Cooper, 1976), pp. 82–88. While such training did occur, I doubt if it was of sufficient scale to warrant the importance these authors attach to it. Certainly, Hobart would have known of any significant work going on prior to his appointment and not only would have told Liddell Hart about it, but also would not have been nearly so negative about going there. Their correspondence, however, indicates this was not the case.

71. "Army Training Memorandum No. 19," dated 5 December 1937, was written before Hobart was appointed DMT, but issued after he took office. "Army Training Memorandum No. 20" was issued on 20 April 1938, "Army Training Memorandum No. 21" was dated 2 December 1938, but written before Hobart left office. General Pile confirmed the time lag in these instances as normal for the period.

72. "Army Training Memorandum No. 21," dated 31 December 1938, paras. 16 and 17.

73. Ibid.

74. Liddell Hart, "Army Reorganization," an unpublished paper dated 16 November 1937 for Hore-Belisha, A4/16, Liddell Hart Papers. Gort expressed agreement with this in a conversation with Liddell Hart on 20 May 1938, "Notes for History," A4/15, Liddell Hart Papers.

75. Martel, *Our Armoured Forces,* p. 378; R. M. Ogorkiewicz, *Armoured Forces,* p. 159. See chart on p. 218.

76. "Army Training Memorandum No. 21," dated 21 December 1938, para. 71, stated that the design of the cruiser tank was "well-advanced" and hoped for production of certain types early the next year.

77. Great Britain, War Office, *The Half Yearly Army List for the Period Ending 31 December,* 38 (London: HMSO, 1939): 65.

78. Conversation between Pile and Liddell Hart on 5 May 1939, "Notes for History," Liddell Hart Papers.

79. Pownall, *Chief of Staff: The Diaries,* pp. 165–66, entry for 17 October 1938; 172–73, entry for 5 December 1938; pp. 173–74, entry for 12 December 1938.

80. Ibid., p. 186, entry for 6 February 1939.

81. Ibid., pp. 188–89, entry for 20 February 1939.

82. The battalions were later redesignated regiments.

83. WO 33/1509, "Report by a Committee Assembled to Consider the Organization for the Mechanized Cavalry and the Royal Tank Corps," 1938, pp. 5–7.

84. Ibid., pp. 20–21.

85. Paul R. Shay, Jr., *British Rearmament in the Thirties* (Princeton, N.J.: Princeton University Press, 1977), pp. 272–73; Pownall, *Chief of Staff: The Diaries,* p. 199, entry for 17 April 1939.

86. Liddell Hart, *Memoirs,* 2:235.

87. M. M. Postan, *British War Production* in *History of the Second World War, United Kingdom Civil Series,* Sir Keith Hancock, gen. ed. (London: HMSO, 1952), pp. 104–05.

88. Conversations between Liddell Hart and Hore-Belisha, 26 July 1939, "Notes for History," Liddell Hart Papers.

89. Conversations between Liddell Hart and "C," 28 August 1939 and Liddell Hart and Broad, 11 July 1939, "Notes for History," Liddell Hart Papers; Peter Chamberlain and Chris Ellis, *British and American Tanks of World War II* (New York: Arco Publishing Company, 1969), p. 56; Liddell Hart, *The Tanks* 2:6.

90. Major L[ionel] F. Ellis, *The War in France and Flanders, 1939–40* in *History of the Second World War, United Kingdom Military Series,* Sir J[ames] R. M. Butler, gen. ed. (London: HMSO, 1953), contains a detailed organization for each of these units, Appendix 1, pp. 369–70. The mechanized cavalry regiments also had 44 light machine gun carriers each.

91. Postan, *British War Production,* pp. 53–54.

92. Ironside, *The Ironside Diaries,* pp. 128–29. This comment is by Colonel MacLeod, who was Ironside's military assistant.

93. Ibid.

94. Ibid., pp. 38–39. Ironside later recalled that, when Hore-Belisha asked him to assume the post of CIGS, he replied, "What the hell for?" Ironside to MacLeod, 1 May 1954, Ironside Papers.

95. Liddell Hart, "Evaluation of Senior Army Officers," an unpublished memorandum dated 8 August 1937, Liddell Hart Papers.

96. Ironside to MacLeod, 11 May 1952, Ironside Papers, III.

97. Hore-Belisha, *The Private Papers of Hore-Belisha,* p. 240.

98. Ibid., p. 240.

99. During the last third of 1939 Great Britain produced 314 tanks to 247 for Germany; for the first third of 1940 it produced 287 to 283 for Germany. Poston, *British War Production,* p. 109.

100. Martel, *Our Armoured Forces,* p. 349.

101. Liddell Hart, *The Tanks,* 2:23. The remaining tanks were all light models.

102. Ironside, *The Ironside Diaries,* p. 400, telegram from Ironside to Hore-Belish, 30 November 1939.

103. F. L. Ellis, *The War in France and Flanders,* pp. 358–69, contains the Order of Battle for the BEF. The divisional cavalry regiments each consisted of twenty-eight light tanks and forty-four lightly armored, tracked machine gun carriers. The army tank battalions consisted of fifty infantry tanks and eight light tanks.

104. Pownall, *Chief of Staff: The Diaries,* p. 243, entry for 12 October 1939.

105. Anonymous, "Foreign News and Views on Tanks," *Royal Tank Corps Journal* 5 (January 1939): 42.

106. Anonymous, "Tanks in Spain," *Royal Tank Corps Journal* 5 (January 1939): 39.

107. For example, see Anonymous, "Tanks in the Attack: French and German Tactical Theories," *Journal of the Royal United Service Institution* 83 (November 1938): 747; Anonymous, "War: Some Recent German Ideas," *Army Quarterly* 39 (October 1939): 34–43; Anonymous, "Foreign Views and News on Tanks," *Royal Tank Corps Journal* 5 (January 1939): 44.

108. Anonymous, "Tank Tactics," *Tank Journal* 6 (July 1939): 29.

109. Anonymous, "The Tank in the 'Deliberate Attack: A Comparison Between Germany and France," *Royal Tank Corps Journal* 5 (January 1939): 34; see also Anonymous, "Tank Tactics," *Tank Journal* 6 (July 1939): 27–29; and Anonymous "Foreign Doctrines on Tank Employment," ibid., pp. 30–31.

110. Anonymous, "Mechanized Warfare," *Royal Tank Corps Journal* 5 (January 1939): 23–26, is an extended review of Eimannsberger's book *Kamfwagenkrieg* (Tank Warfare), first published in 1934. No other German author received such extensive treatment. In addition, Eimannsberger received references in several of the articles to which reference has already been made.

111. Ibid., p. 25.

112. Anonymous, "Tanks in the Attack: French and German Tactical Theories," *Journal of the Royal United Service Institution,* 747; Anonymous, "German Ideas about Tanks," *Army Quarterly* 40 (April 1940): 54.

113. WO 190/723, "Report on German Armour Tactics," an undated intelligence report by the British Military Attaché. Gort thought it of sufficient interest to forward it to Hore-Belisha with a note, dated 21 November 1938.

114. WO 197/25, "Report on the Employment of Armoured Forces," Minute 3, dated October 1939.

115. Heinz Guderian, *Panzer Leader* (New York: E. P. Dutton and Co., 1952), p. 20.

116. Brian Bond, *Liddell Hart: A Study of His Military Thought* (New Brunswick, N.J.: Rutgers University Press, 1977), pp. 215–235. Specifically, Bond has concluded that (1) Fuller and Liddell Hart deserve equal recognition as advocates of armored war in Germany, and (2) that Liddell Hart's most important contributions came before the formation of the first Panzer divisions and lay in his advocacy of deep strategic penetrations behind enemy lines by massed armored formations containing mechanized infantry and other supporting arms. Ibid., p. 235.

117. This survey of German military thought is based on Larry H. Addington, *The Blitzkrieg Era and the German General Staff, 1865–1941* (New Brunswick, N.J.: Rutgets University Press), pp. 3–42. These strategic theories were also affected by material conditions such as the state of the German economy, as Professor Addington points out very well. For the purposes of this study, however, a summary of the evolution of these principles is sufficient.

• 8 •

Epilogue:
The Test of Battle

Introduction

BETWEEN the opening of the battle for France on May 10, 1940, and the first battle of El Alamein on July 1, 1942, the British Army suffered a succession of defeats at the hands of the German Army that were hardly relieved by its successes against the Italians in Africa during the winter of 1940–41. The reasons for these failures are many and complex. They include the personal weaknesses of some field commanders, the lack of sufficient equipment and often the inadequacy of what it had, the poor state of training of many units for the tasks before them, and, in the case of the campaign in France, the considerable weaknesses of the French Army. Underlying all of these problems, however, was the British Army's continued adherence to the strategy of attrition and the vulnerability of that strategy to blitzkrieg warfare. Indeed, if the French and North African campaigns are examined from this perspective, the plans and decisions of British commanders assume a degree of logic they do not otherwise have and the reasons for their failures stand out more clearly.

France, 1940

Anglo-French plans for the defense of France in the spring of 1940 were as simple as they were conventional. They assumed the Germans would attempt a repetition of the Schlieffen Plan (this time to include Holland) and planned to counter it by moving into Belgium. From the point where the Meuse River crosses the French border to the coast, they would move in an extended line (much like a swinging door) to take up positions as far forward as the rate of the German advance and the reaction of Belgium and Holland allowed.[1] At this point they envisaged a major battle in which the German advance would be halted and contained. Planning beyond these opening moves could not, of course, be very specific, but in general terms the Allies planned to remain on the strategic defensive while Great Britain built up its army until the Allies had sufficient strength to go on the offensive. At that point—which could be no earlier than mid-1942 because of shortages of equipment[2]—they would then launch a major drive to push the Germans back. The role of armor in this grand plan was a direct application of the cavalry concept of tank warfare. In the main, the French and British both planned to employ tanks for reconnaissance, screening, direct infantry support, and containment of German armored formations that might penetrate their front. Accordingly, their advance into Belgium was to be preceded by mechanized cavalry regiments that would rush ahead to seize key terrain features until the slower infantry divisions could move up, while the army tank battalions would work directly with the infantry. The First Armoured Divison, when it arrived, would be used as a mobile reserve to block and, if possible, destroy any German armor that might break through; until that time, the army-tank battalions would perform that function.[3] The similarity between these plans and the strategic doctrines of World War I is obvious. All of the elements of the Napoleonic model of war were present, only slightly modified by the increased range and speed afforded by the mechanization of the army's transport. The concept of operations reflected the three phases—advance to contact, main battle, and pursuit—that they had been taught to expect. The role of armor likewise fit into this pattern. Tanks were to take the place of the old horsed cavalry and, like them, were divided into two elements: fast-moving light forces to operate on the periphery of the main body of the army to locate the enemy and cover the movement of their own forces; and slower-moving heavy forces to take part in the main battle and work closely with the infantry and artillery in offensive action. As a result, while the British and, to a lesser degree, the French had conceded tactical independence to their armored forces, they continued to see them as specialized units rather than versatile, self-contained combat formations and to tie them strategically to the slower arms, restricting their radius of action. Furthermore, by advancing on an extended line as they planned, the Allies insured not only that their forces would lack sufficient depth to counter a deep strategic penetration by German armor, but also that they would not be sufficiently concentrated to deliver a decisive blow of their own.

The final German plan for the 1940 campaign, given the code name *Sichelschnitt* ("the cut of the sickle"), was far more audacious than the Allies ever dreamed. It called for a massive armored thrust aimed not at tactical objectives such as the isolation of a brigade or division, but at the entire left wing of the Allied forces. Its objective was not to penetrate the Allied battle area, but to drive deeply into the interior of France, isolate the entire left wing of the Allied army, and pin it down until the main body of the

German Army could come up and deliver the final blow. Like all ambitious plans, it was fraught with great risks, and its success depended not on numbers, but on surprise and, above all, speed. This was the means by which the Germans would prevent the Allied from delivering an effective counterattack. Such a plan could succeed only if the German armor was seen as more than revitalized cavalry, and if it was made strategically independent of the remainder of the army, and this is exactly what the German Army had done.[4]

The organization of the respective armored forces was a further indication of the differences between the manner in which each side planned to use its tanks. As befitted the cavalry concept of tank warfare, the Allied armored forces contained at most only a very small proportion of supporting units from the other combat arms. The army-tank brigades had only two tank battalions with no infantry whatsoever. The mechanized cavalry regiments, four of which were grouped together into light armored reconnaissance brigades of two regiments each, likewise had no infantry. Only the First Armoured Division (which was not present) contained support elements from other arms, but its tank-infantry ratio of three to one was much too low. By comparison, the German Panzer divisions contained four motorized infantry battalions to four tank battalions (plus supporting units), making them truly effective all-around combat forces.[5] The higher organization of the armored forces further reflected these differences. Of the 3,450 Allied tanks at the beginning of May, only about 450 were organized into armored divisions. The arrival of the First Armoured Division would have increased this significantly, but, even so, it would have meant that only twenty percent of the Allied armor was in division-scale units. The Germans, on the other hand, organized their armored forces to reflect their strategic independence. All of the German tanks were in the Panzer divisions, and, for their main thrust, seven of these divisions were further concentrated into a Panzer group of two corps (totalling five divisions) and a separate corps. The entire force totalled 1,806 tanks and would be further strengthened by two more armored divisions during the drive to the sea. Against this, the largest organized armored unit the Allies could muster was only 150 tanks.[6]

The course of the campaign followed logically from the concepts with which the contending forces entered the battle. The Allies moved to their positions on the Dyle in a conventional manner, the British facing no opposition whatsoever, and prepared to meet the Germans head-on. Meanwhile, the Germans came through the weakly defended Ardennes and sent their armor dashing to the sea. In only six days, May 15–20, the Panzers advanced some 160 miles behind the Allied lines, and on the last day the leading units advanced sixty miles. The obvious solution to this threat was for the Allies to counter with a pincer attack in the gap between the Panzers and the more slowly moving infantry. Such an attempt was made on May 21 in the region of the old World War I battlefields north of the Somme. The plan was for the French to attack north through Albert and the British south from Arras, but the operation never got off the ground. The French attack never took place and the British (commanded, incidentally, by Martel) could muster only two infantry battalions, a tank force of fifty-eight Mark I and sixteen Mark II infantry tanks, and some artillery. After a slight advance they were halted, and, although the Germans experienced considerable anxiety for a few hours, they were never in danger of being isolated.[7] From the Allied point of view, the problem that proved insuperable was the time factor. Strategic doctrines not only determine how commanders will react to a situation, but they also determine how fast they will react. For the Allied commanders and staffs in May 1940, the strategy of blitzkrieg warfare demanded movement at a pace they could not meet. Trained to expect that the main blow would come only after a long struggle of attrition between the main bodies of the opposing armies, they could not plan and execute the rapid redeployment of their units the new strategy demanded. Numbed and shaken, they could only stumble back to the channel coast. General Ironside summed up the situation succinctly when, after visiting the French First Army on May 20, he commented: "Ready to be slaughtered. Defeated at the head without casualties."[8] He was right. Given the size of the forces employed, losses on both sides were extremely light. Moreover, despite the presence of some 6,000 tanks on both sides, there was no "battle" as that word was traditionally used. General Pownall, the BEF chief of staff, undoubtedly expressed the frustration of many believers in the strategy of attrition when he confided to his diary on the day after the Germans reached the coast: "It is a hopeless position and too absurd because we haven't yet encountered the real German Army; this is only their [armoured] cavalry that is sending us reeling."[9]

North Africa, 1940–42

Stretching some 600 miles from El Alamein in the east to El Agheila in the west, the North African theater from 1940 to 1942 offered an ideal terrain for the practice of blitzkrieg. Rommel, as is well known, practiced such warfare, but the British, despite the lessons of the French campaign and their growing experience in the desert, never did. This is not to say that they learned nothing from their defeats. In the aftermath of Dunkirk, they tripled the number of planned armored divisions to nine and for the first time assigned more tanks to these units than for infantry support. They also reorganized the armored divisions, providing them with additional infantry support to make them more effective all-around combat units.[10] But they continued to seek victory in set-piece battles waged in accordance with the principles of the strategy of attrition. And when, in the end, they won, it was not because they had learned how to fight in accordance with the principles of blitzkrieg, but because conditions had changed so that those principles could no longer be employed against them.

The African campaign opened in December 1940 with a daring British offensive that, at first sight, seems to belie the foregoing analysis. Under the direct command of Lieutenant General Richard N. O'Connor, a British force of only 36,000 men, 275 tanks, and 150 guns attacked an Italian army totaling 250,000 men, 300 tanks, and 400 guns. Spearheaded by the Seventh Armoured Division, which Hobart had trained on the eve of the war, the British virtually destroyed the Italian forces in two months and advanced 500 miles. But was the offensive based on the principles of the strategy of armored war? The answer, on close analysis, is no. Rather, the British offensive was a series of limited operations initiated only to stun the enemy, and it achieved success only through brilliant opportunism on the part of the British and incredible meekness on the part of the Italians. Having advanced some fifty miles into Egypt the previous September, the Italians had scattered their troops into a series of fortified positions well beyond supporting distance of each other. By December, it was apparent that they had little desire to advance further, and the British decided to seize the initiative with their meager forces by launching limited raids to seize the most vulnerable of these positions. The plan was quite simple: using the mobility of the Seventh Armoured Division, they would first isolate a position, then, having concentrated a sufficient force of infantry and powerful Mark II infantry tanks (in the newly arrived Seventh Royal Tank Regiment), they would take the position by assault. The operation was immediately successful. Poorly trained and led and unable to cope with the British heavy armor, one Italian garrison after another quickly succumbed. In less then a week, O'Connor cleared Egypt and, expanding the operation, attacked into Lybia, employing the same tactics and meeting with the same success. After a three-week delay because of a change of units, the British advance resumed. By early February, the Italians were in full retreat from western Lybia only to be blocked at Beda Fomm after another daring desert sweep by the Seventh Armoured Division and forced to surrender. At a cost of under 2,000 casualties, O'Connor had captured some 130,000 prisoners, 380 tanks, and 845 guns.[11]

There can be no doubt that this victory was a remarkable achievement, but the fact remains that it lacked the ingredient—vital in the theories of Fuller and Liddell Hart—of strategic paralysis. The task of the Seventh Armoured Division was never to strike at the enemy's brain, but simply to prevent him from extricating himself from a series of traps into which he had freely entered. Nor did the movements of the division ever, of their own accord, force any of the Italian fortified positions to collapse; that came about only as a result of a Cambria-like tank-infantry assault, in which the vital element was the Italians' inability to destroy the heavy British infantry tanks. Only at Beda Fomm did this pattern vary, but the principle was not essentially different. It was, to repeat, a brilliant campaign, but it was not an application of the theory of armored warfare.[12]

The immediate consequence of O'Connor's victories was to prompt Hitler to dispatch Rommel and a small Panzer and motorized-infantry force to Africa. The British, in turn, were greatly weakened by Churchill's demand that part of their forces be dispatched to Greece, and they lost their most experienced combat units as a result. When Rommel, realizing the British weakness, attacked in March, they were sent reeling back to the Egyptian frontier, managing to hold only the isolated garrison at Tobruk. Even worse, O'Connor was captured during the German advance. Thus, by the late spring of 1941, the British were not simply back to their starting position of the previous November, but without their most battle-tested units and commander.

The effects of these weaknesses were revealed in the fighting along the frontier from May to November 1941. During this period the British launched two major offensives, Operation Battleaxe in June and Operation Crusader in November. Both revealed that the army had lost the sharp fighting edge it had possessed the previous year. More important for our purposes, however, are the plans upon which the operations were based. In both instances, they were designed to bring the Germans to battle under conditions that would enable the British to destroy them. Operation Battleaxe was a straightforward frontal assault on the German line: infantry supported by infantry tanks would attack the fixed German defenses along the coast and the Seventh Armored Division would strike inland to seek out the German armor. According to the plan, when the coastal attacks had secured the German defensive positions and the German forward armor had been destroyed, the British would relieve Tobruk and complete the destruction of the German forces. Unfortunately for the British, German tactical skill was far superior to their own and the offensive quickly broke down, with heavy losses.[13]

Operation Crusader was a far larger and more complex operation. Reinforced and reorganized, the British forces, now designated the Eighth Army, consisted of two full corps with some 500 cruiser and light tanks and 200 infantry tanks against somewhat fewer than 400 German and Italian tanks. The British once again planned to attack with two wings, but this time they hoped to fix the German forces along the coast with their infantry and sweep around the desert flank with their armor. Once in the German rear, they intended to crush the German armor in a decisive tank battle and then move on to Tobruk. Unfortunately, once the British armor penetrated behind German lines, they were unable to locate the German armor. Somewhat bewildered, they dissipated their strength in a series of uncoordinated actions. For three weeks, the battle flowed back and forth until the Eighth Army's superior numbers eventually wore Rommel's forces down, and he was forced to retreat to El Aghleila. Crusader was a British victory, but it was a costly one.[14] The process of attrition had worked for the British, but as Fuller and Liddell Hart had so often maintained, the price of such victories could easily pave the way for later defeat. In the words of the *Official History,* Crusader had "consumed so many of their [the Eighth Army's] resources . . . by the time they reached the western end of Cyrenaria their blow was spent."[15] In January 1942, Rommel struck again, driving the Eighth Army first back to Gazala and then, in June, to El Alamein, the last defensible position before Alexandria and the Nile delta. In July and again in August and September, he attempted to break through the British positions; but his extended supply lines, the geographical limitations of the El Alamein position, which greatly restricted his maneuvering space, and the growing tactical skills of the Eighth Army halted him. By the end of September, the British were once again ready to go on the offense.

The second battle of El Alamein was the watershed for the British Army in World War II. It was the army's first clearly recognizable victory against the Germans. It signalled the emergence of General (later Field Marshal the Viscount) Montgomery as the army's foremost field commander. But most important, it was a victorious battle planned and executed in accordance with the principles of the strategy of attrition, and, because of that, it insured that those principles would never again be challanged during the war. As Montgomery told his senior officers on the eve of battle:

> Methodical progress, destroy enemy part by part slowly and surely. Shoot tanks and shoot Germans. He cannot last a long battle; we can. We must therefore keep at it hard; no unit commander must relax the pressure. Organize ahead for a "dogfight" of a week. Whole affair about 10 days. Don't expect spectacular results too soon.[16]

Yet Montgomery's offensive was not a repetition of the methods used on the Western Front in the previous war. As a master of the set-piece battle, he employed his forces skillfully, pressing his attacks aggressively when they offered a chance of success, shifting them quickly when they did not. His success at El Alamein was the result in part of this skill, but it was also the result of his material superiority and of Hitler's refusal to allow Rommel to retreat until it was too late. Forced to defend an advanced position far from his source of supply, Rommel had no opportunity to practice the highly mobile form of war for which he was justly famous. The British victory, in short, was not the result of the fact that they had learned to overcome the challenge of blitzkrieg warfare; it was the result of the fact that the Germans were no longer in a position to employ it. The pattern for the remainder of the war had been set. As Larry Addington has pointed out in his excellent study of the blitzkrieg, the fine cutting edge of the

German Army upon which their success depended had always been very small. As the war expanded, German industry was unable to expand or even maintain that edge. As a result, the German Army steadily lost its mobility as its enemies increased theirs. In the end, it was forced to fight everywhere as Rommel had been forced to fight at El Alamein—in great struggles of attrition against foes vastly superior in both human and material resources.[17]

Given this fact, it is obvious that the British Army's success later in the war does not mean that it had necessarily found the best solution to the questions of mechanization and strategic doctrine that had dominated so much of its attention during the interwar years. This study has shown that both these questions and the army's reaction to them were, in fact, very different; and that while the army was not nearly so conservative regarding the former as it has sometimes been accused of being, it was rigidly fixed regarding the latter. It may not be a terribly profound lesson of history, but it is one worth remembering in an age of constantly changing technology: that strategic doctrines are not unchanging articles of faith, but rather, are guides to action arising out of civil and military conditions of a particular time; and that to adopt new weapons and merely fit them into established strategic doctrines without any thought as to how they might change those doctrines is both irresponsible and dangerous. This was the failure of the British Army during the interwar years, a failure for which it paid a heavy price between 1939 and 1942.

As for the early band of British tank advocates, the war seemed to pass them by as it had the ideas (but not the weapon) they urged the army to adopt. None of them held a major field command during the war. Lindsay and Broad retired early in the war because of age, Martel served in various staff positions, and Pile assumed command of all of the antiaircraft defenses in Great Britain. Only Hobart continued to work directly with tanks throughout the war, and, after the Normandy invasion, commanded the Seventy-ninth Armoured Division. This, however, was composed entirely of special armored vehicles (flame-throwing tanks and bridge-laying tanks, for example) and his was largely an administrative post. His real ability as a combat leader was never utilized to its full potential.

For Fuller and Liddell Hart, the war also spelled disappointment, the former for his fascist views and the latter for the misunderstanding caused by his writings during the winter of 1939–40. With the passing of years, however, both men began to receive the recognition they deserved. In 1963, they were jointly awarded the Gold Medal for outstanding contributions in the field of military writing by the Royal United Service Institution, and, in 1966, Liddell Hart was knighted. Both also continued to be prolific writers. At the age of eighty-three, Fuller published one of his most important analyses of modern warfare, entitled *The Conduct of War, 1789–1961*. With all of its prejudices—and Fuller always had many—it is one of the most thought-provoking studies of the subject in recent years. He died in 1966 at the age of eighty-eight. Liddell Hart survived his close friend and associate by only little more than three years. Like Fuller, he continued writing until his death and completed reading the proofs of his last book, *History of the Second World War,* only a few weeks before he died in January 1970. In many respects—although certainly not in the traditional one—he will be justly remembered as one of the great captains of history."

Notes

1. The exact line could not be determined beforehand because Belgian and Dutch authorities, hoping to remain neutral, refused to commit themselves to cooperation with England and France. As a result, Allied plans called for an advance either to the Escaut River if the Belgians did not join them or to the Dyle River if they did. Sir J[ames] R. M. Butler, *Grand Strategy,* vol. 2: *September 1939–June 1941,* in *History of the Second World War, United Kingdom Military Series,* Sir J[ames] R. M. Butler, gen. ed. (London: HMSO, 1957), pp. 157–64

2. Butler, *Grand Strategy,* 2: 28.

3. For the French view, see Alistair Horne, *To Lose a Battle* (London: Macmillan, 1969), p. 64. The British view is contained in WO 173/13, "Employment of Armour in the B. E. F. 1939–1940" dated September 1939.

4. Horne, *To Lose a Battle,* p. 138.

5. Ogorkiewicz, *Armoured Forces* (London: Arms and Armor Press, 1970), pp. 73–75. This organization held for the original six Panzer divisions. The remaining four were light divisions converted after the Polish campaign. Three of them had three tank battalions and one had two.

6. Horne, *To Lose a Battle,* end papers. The main German thrust was also supported by four motorized infantry divisions.

7. Ibid., pp. 444–45.

8. Edmund Ironside, *The Ironside Diaries, 1937–1940,* Macleod And Denis Kelley, eds. (London: T. A. Constable Ltd., 1962), p. 321.

9. Henry Pownall, Chief of Staff: *The Diaries of Lieutenant Sir Henry Pownall,* 2 vols., B. Bond, ed., (London: Anchor Books, 1973), 1:327.

10. Ogorkiewicz, *Armoured Forces,* p. 60.

11. Correlli Barnett, *The Desert Generals,* (New York: Viking, 1961), pp. 19–63; F. G. Jackson, *The Battle for North Africa* (N.Y.: Mason, 1975), pp. 36–80.

12. This conclusion closely parallels that of Field Marshal Lord Michael Carver, *The Apostles of Mobility: The Theory and Practice of Armoured Warfare* (London: Holmes & Meier, 1979), pp. 69–70.

13. Barnett, *The Desert Generals, pp. 67–75; Jackson, The Battle for North Africa,* pp. 155–167.

14. Barnett, *The Desert Generals,* pp. 90–114; Jackson, *The Battle for North Africa,* pp. 180–229.

15. Major General I[an] S. O. Playfair, *The Mediterranean and Middle East,* vol. III: *British Fortunes Reach Their Lowest Ebb,* in *History of the Second World War, United Kingdom Military Series,* Sir J[ames] R. M. Butler, gen. ed. (London: HMSO, 1960), p. 96.

16. B. L. Montgomery, *The Memoirs of Field Marshal the Viscount Montgomery of El Alamein* (Cleveland and New York: The World Publishing Company, 1958), p. 115.

17. Larry Addington, *The Blitzkrieg Era and the German General Staff, 1865–1941* (New Brunswick, N. J.: Rutgers University Press, 1971), pp. 213–17.

Bibliography

Manuscript Collections and Public Records

Public Records Office

War Office

WO 32/2561, Minutes of Fortnightly Meetings, Army Council, 1933.

WO 32/2820, Proposals for Initiation of the Experimental Mechanical Force, 1927.

WO 32/2821, Proposed Establishment of the Experimental Mechanical Force, 1927.

WO 32/2822, Movement of Units in Connection with Experimental Mechanical Force, 1927.

WO 32/2823, Distribution and Strength of the British Army, 1927.

WO 32/2824, Disarmament—Views of Foch, 1928.

WO 32/2825, Experimental Formations for 1929.

WO 32/2826, The Mobile Division and Mechanization of the Cavalry.

WO 32/2828, Experimental Armoured Force, Report for 1928.

WO 32/2838, Future of the Royal Horse and Light Artillery, 1928.

WO 32/2840, Mechanization of Division Artillery, 1928.

WO 32/2841, Cavalry Committee Interim Report, 1926.

WO 32/2842, Cavalry Committee Final Report, 1926.

WO 32/2845, Future Organization of the Cavalry, 1927–28.

WO 32/2846, Reducing Money for Cavalry, 1927.

WO 32/2847, Introduction of the Mobile Division, 1934–36.

WO 32/2852, The Tank Situation, 1932.

WO 32/3878, Object of the Imperial Defence College, 1935.

WO 32/4005, Report of the Committee on the Value of the Horse in the Training of Officers, 1935–40.

WO 32/4195, Proceedings of the Informal Army Council Meetings from July 1937 to February 1938.

WO 32/4441, The Tank Situation, Progress Reports, 1936–40.

WO 32/4580, Proposal of the Secretary of State for War to Promote Rapid Conduct of Army Council Business, 1937.

WO 32/4612, Future Reorganization, 1935.

WO 32/4633, The Suggested Retention of the Scots Greys as Horsed Cavalry, 1936–37.

WO 32/4840, Report on Higher Education for War and Reorganization of the Senior Division at the Staff College, 1927.

WO 32/5959, Reduction of the Cavalry Regiments, 1921.

WO 32/9405, Delegation of Powers—Army Council and Secretary of State for War, 1937.

WO 33/1015, Interim Report of the Committee on Sea Transport and Tanks, 1922.

WO 33/1023, Composition of the British Expeditionary Force for Small Wars, 1923–24.

WO 33/1058, Composition of the British Expeditionary Force, 1924.

WO 33/1183, Supplement to Mechanized and Armoured Formations, 1929.

WO 33/1187, Composition of the British Expeditionary Force, 1930.

WO 33/1207, Expeditionary Force Tables for the Regular Expeditionary Force, 1930.

WO 33/1237, Mechanical Warfare Board: First Report, 1927–28.

WO 33/1238, Expeditionary Force Tables for the Regular Expeditionary Force, 1931–33.

WO 33/1243, Mechanical Warfare Board: Second Report, 1929–30.

WO 33/1258, Modern Formations Supplement, 1931.

WO 33/1283, Mechanical Warfare Board: Third Report, 1930–31.

WO 33/1297, Report of the Committee on the Lessons of the Great War, 1932.

WO 33/1303, Expeditionary Force Committee: Interim Report, 1932.

WO 33/1307, Mechanical Warfare Board: Fourth Report, 1931–32.

WO 33/1338, Mechanical Warfare Board: Fifth Report, 1932–33.

WO 33/1341, Composition of the Field Force and Remaining Formations and Units Not Allotted to the Field Force in the Western Plan, 1934.

WO 33/1367, First Report of the Mechanization Board, 1934.

WO 33/1377, Second Interim Report of the Field Force Committee, 1935.

WO 33/1404, Second Report of the Mechanization Board, 1935.

WO 33/1415, The Intermediate Contingent Field Force with Amendments, 1935.

WO 33/1434, Third Interim Report of the Field Force Committee, 1936.

WO 33/1459, Third Report of the Mechanization Board, 1936.

WO 33/1472, Fourth Interim Report of the Field Force Committee, 1937.

WO 33/1491, Second Report of the Committee on the Supply of Army Officers.

WO 33/1502, Memorandum by the Secretary of State of War on the Organization of the Army for Its Role in War.

WO 33/1505, Fourth Report of the Mechanization Board, 1937.

WO 33/1509, Report by a Committee Assembled to Consider the Organization of the Mechanized Cavalry and the Royal Tank Corps, 1938.

WO 163/25–43, Minutes of the Proceedings of, and Precis Prepared for the Army Council, 1920–37.

WO 173/13, Employment of Armour in the B. E. F., 1939–40.

WO 190/637, Appreciation of German Armour Tactics, 1938.

WO 190/723, German Tank Tactics, 1938.

WO 190/736, Light and Armoured Divisions of the German Army, 1939.

WO 197/25, Report on the German Employment of Armoured Forces, 1939.

Center for Military Archives, Kings College, University of London

Lieutenant General Sir Charles Broad Papers

Major General J.F.C. Fuller Papers

Field Marshal Sir Edmund Ironside Papers

Field Marshal Sir Archibald A. Montgomery-Massingberd Papers

Brigadier H. R. Sandilands Papers

Liddell Hart Archives, States House, Medmenham, Marlow, Buckinghamshire, England

Major General Sir Percy Hobart Papers

Sir Basil Liddell Hart Papers

Major General John Lindsay Papers

Special Collection Department, Rutgers University Library, New Brunswick, New Jersey

Major General J. F. C. Fuller Papers.

Newspapers

Daily Telegraph. 31 August, 28 September, 23 November 1925.

19 February, 31 March, 26 April, 19 July, 28 August, 1 September, 24 September, 14 December 1926.

22 April, 12 May, 20 August, 31 August, 10 September, 16–17 September, 20 September, 24 September, 27 September 1927.

12 April, 8 May, 6 July, 13 July, 25 August, 25 September, 26 September 1928.

12 March, 13 March, 26 September, 30 September 1929.

21 January, 13 March, 29 May, 9 September, 25 September, 26 September 1930.

23 August, 25 August, 1–3 September, 5 September, 10 September, 13 September, 14–17 September 1931.

23 August 1932

20 February, 28 March, 30 March, 1 April, 3 October, 11 November 1933.

28 August, 24 September, 15 November 1934.

14 March 1935.

The Times. 14 March, 18 March, 8 August, 23 August, 27 August, 18 September, 19 September, 27 December 1935.

3 September, 4 September, 2 November 1936.

5 March, 22 September 1937.

Personal Interviews

Lieutenant Sir Charles Broad at his home, Beaminster, Dorset, 19–20 August 1970.
General Sir Frederick Pile at the Cavalry Club, London, 6 August 1971.

Published Sources: Primary

Books

Clausewitz, Karl von. *A Short Guide to Clausewitz on War.* Edited by Roger A. Leonard. New York: Capricorn Books, 1968.

———. *On War.* 3 vols. Translated by J. J. Graham, with notes and introduction by F. N. Maude. London: Kegan Paul, Trench, Trubner & Co., Ltd., 1918.

Cooper, Duff. *Old Men Forget.* New York: E. P. Dutton & Co., Inc., 1954.

Foch, Ferdinand. *The Principles of War.* Translated by J. de Morinni. New York: The H. K. Fly Co., 1918.

Fuller, J. F. C. *Armoured Warfare.* Harrisburg, Pa.: The Military Service Publishing Co., 1943.

———. *The Army in My Time.* London: Rich & Cowan, Ltd., 1935.

———. *British Light Infantry in the Eighteenth Century.* London: Hutchinson & Co., 1925.

———. *The Conduct of War: 1789–1961.* New Brunswick, N.J.: Rutgers University Press, 1961.

———. *The Dragon's Teeth: A Study of War and Peace.* London: Constable & Co., Ltd., 1932.

———. *The First of the League Wars: Its Lessons and Omens.* London: Eyre & Spottiswoode, 1936.

———. *The Foundations of the Science of War.* London: Hutchinson & Co., Ltd., 1926.

———. *On Future War.* London: Sifton Praed & Co., Ltd., 1928.

———. *Generalship: Its Diseases and Their Cure.* Harrisburg, Pa.: Military Service Publishing Co., 1936.

———. *The Generalship of Ulysses S. Grant.* New York: Dodd, Mead & Co., 1929.

———. *Grant and Lee: A Study in Personality and Generalship.* Bloomington, Ind.: Indiana University Press, 1957 (reprint of 1933 edition).

———. *Imperial Defence: 1585–1914.* London: Sifton Praed & Co., Ltd., 1926.

———. *Lectures on Field Service Regulations II.* London: Sifton Praed & Co., Ltd., 1931.

———. *Memoirs of an Unconventional Soldier.* London: Ivor Nicholson & Watson, Ltd., 1936.

———. *Sir John Moore's System of Training.* London: Hutchinson & Co., 1925.

———. *The Reformation of War.* New York: E. P. Dutton & Co., 1923.

———. *Tanks in the Great War.* London: John Murray, 1920.

———. *Toward Armageddon: The Defense Problem and Its Solution.* London: Lovat Dickson, Ltd., 1937.

———. *War and Western Civilization: 1832–1932.* London: Duckworth, 1932.

Germains, Victor. *The Mechanization of War.* London: Praed & Co., Ltd., 1928.

Great Britain. Parliament. *Parliamentary Debates* (Commons), 3rd Series, 89 (1921).

———. *Parliamentary Debates* (Commons), 3rd Series, 193 (1926).

Great Britain. War Office. *Field Service Regulations.* Part 1, *Operations.* London: HMSO, 1909 (reprinted, with amendments, 1912).

———. *Field Service Regulations.* Vol. 2, *Operations.* London: HMSO, 1920.

———. *Field Service Regulations.* Vol. 2, *Operations.* London: HMSO, 1924.

———. *Field Service Regulations.* Vol. 2, *Operations.* London: HMSO, 1929.

———. *Field Service Regulations.* Vol. 2, *Operations.* London; HMSO, 1935.

———. *Field Service Regulations.* Vol. 3, *Higher Operations.* London: HMSO, 1935.

———. *The Half-Yearly Army List for the Period Ending 31 December 1922.* London: HMSO, 1923.

———. *The Half-Yearly Army List for the Period Ending 31 December 1938.* London: HMSO, 1939.

———. *Mechanized and Armoured Formations.* London: HMSO, 1929.

———. *The Monthly Army List for August 1932.* London: HMSO, 1932.

———. *The War Office List.* London: HMSO, 1928.

Guderian, Heinz. *Panzer Leader.* New York: E. P. Dutton & Co., 1952; London: Michael Joseph, Ltd., 1952.

Haig, Sir Douglas. *The Private Papers of Douglas Haig.* Edited by Robert Blake. London: Eyre & Spottiswoode, 1952.

———. *Sir Douglas Haig's Dispatches.* Edited by Lieutenant Colonel J. H. Boraston. London: J. M. Dent & Sons, Ltd., 1920.

Hankey, Maurice. *The Supreme Command, 1914–1918.* 2 vols. London: George Allen & Unwin, Ltd., 1961.

Hore-Belisha. *The Private Papers of Hore-Belisha.* Edited by R. J. Minney. Garden City, N.Y.: Doubleday & Co., Inc., 1961.

Ironside, Edmund. *The Ironside Diaries: 1937–1940.* Edited by Colonel Roderick MacLeod and Denis Kelly. London: T. & A. Constable, Ltd., 1962.

Liddell Hart, B. H. *The British Way in Warfare.* London: Faber & Faber, Ltd., 1932.

———. *The Decisive Wars of History: A Study in Strategy.* London: G. Bell & Sons, Ltd., 1929.

———. *The Defence of Britain.* New York: Random House, 1939.

———. *Europe in Arms.* New York: Random House, 1937.

———. *Foch: The Man of Orleans.* Boston: Little, Brown & Co., 1932.

———. *The Future of Infantry.* London: Faber & Faber, Ltd., 1933.

———. *The Ghost of Napoleon.* New Haven, Conn.: Yale University Press, 1935.

———. *Great Captains Unveiled.* Boston: Little, Brown & Co., 1927.

———. *A Greater Than Napoleon: Scipio Africanus.* London: William Blackwood & Sons, Ltd., 1927.

———. *Memoirs.* 2 vols. London: Cassell, 1965.

———. *Paris; or the Future of War.* London: Kean Paul, Trench, Trubner & Co., Ltd., 1925.

———. *Reputations: Ten Years After.* Boston: Little, Brown & Co., 1928.

———. *The Remaking of Modern Armies.* Boston: Little, Brown & Co., 1928.

———. *The Revolution in Warfare.* New Haven Conn.: Yale University Press, 1947.

———. *A Science of Infantry Tactics Simplified.* London: William Clowes & Sons, Ltd., 1926.

———. *Sherman.* New York: Praeger Paperbacks, 1960 (reprint of 1929 edition).

———. *The Tanks.* 2 vols. New York: Frederick A. Praeger, 1959.

———. *Thoughts on War.* London: Faber & Faber, Ltd., 1944.

———. *Through the Fog of War.* New York: Random House, 1938.

———. *The War in Outline.* New York: Little, Brown & Co., 1930.

———. *When Britain Goes to War.* London: Faber & Faber, Ltd., 1935.

Martel, Gifford L. Q. *Our Armoured Forces.* London: Faber & Faber, 1946.

Montgomery, B. L. *The Memoirs of Field Marshal the Viscount Montgomery of Alamein.* Cleveland and New York: The World Publishing Company, 1959.

Montgomery-Massingberd, Field Marshal Sir Archibald. *The Autobiography of a Gunner.* Privately published by the author's family, 1946.

Pownall, Sir Henry. *Chief of Staff: The Diaries of Lieutenant General Sir Henry Pownall,* Vol. 1, *1933–1940.* Edited by B. Bond. London: Archon Books, 1973.

Robertson, Field Marshal Sir William. *From Private to Field Marshal.* New York: Houghton Mifflin Co., 1921.

Rowan-Robinson, H. *Some Aspects of Mechanization.* London: William Dowes and Sons, Ltd., 1928.

Swinton, Major General Sir Ernest D. *Eyewitness.* Garden City, N.Y.: Doubleday, Doran & Co., Inc., 1933.

Articles

Anonymous. "A Horse! A Horse! My Kingdom for a Horse!" *Army Quarterly* 23 (April 1934): 150–53.

———. "Army Reorganization." *Journal of the Royal United Service Institution* 63 (May 1918): 286–300.

———. "Auream Quisque Mediocritatem DILIGET." *Army Quarterly* 5 (October 1922): 122–29.

———. "The British Campaign in the West." *Army Quarterly,* pt. 1: 5 (January 1923): 314–30; pt. 2, 6 (April 1923): 44–56.

———. "The Employment of a Tank Brigade." *Royal Tank Corps Journal* 4 (July 1938): 4–8.

———. "Employment of Tanks on the Indian Frontier." *Royal Tank Corps Journal* 2 (July 1937): 9–12.

———. "The 1st Tank Brigade." *Royal Tank Corps Journal* 3 (January 1938): 2–4.

———. "Foreign Doctrines on Tank Employment." *The Tank Journal* 6 (July 1939): 30–32.

———. "The Foreign Press." *Royal Tank Corps Journal* 3 (January 1938): 42–44.

———. "Foreign Tank News and Views." *Tank Journal* 6 (July 1939): 45–50.

———. "Foreign Views and News." *Royal Tank Corps Journal* 4 (July 1938): 58–63.

———. "Foreign Views and News on Tanks." *Royal Tank Corps Journal* 4 (January 1939): 42–48.

———. "The Function of the Mobile Division." *Royal Tank Corps Journal* 1 (January 1937): 48–51.

———. "General Joffre's Plan." *Journal of the Royal United Service Institution* 65 (May 1920): 380–87.

———. "A German Account of the British Offensive of August 1918." *Army Quarterly* 6 (April 1923): 11–16. (From General von Zwehl, *Die Schlachten in Sommer 1918*.)

———. "German Ideas about Tanks." *Army Quarterly* 40 (April 1940): 41–55.

———. "The 'I' Tank, Another Infantryman's Point of View." *Royal Tank Corps Journal* 2 (July 1937): 28–29.

———. "Mechanized Warfare." *Royal Tank Corps Journal* 5 (January 1939): 23–26.

———. "The Mobile Division of the Future." *Cavalry Journal* 14 (April 1924): 155–64.

———. "Notes on Co-operation Between Cavalry and Tanks." *Cavalry Journal* 13 (July 1923): 317–23.

———. "The Organization and Administration of the Tank Corps During the Great War, 1916–1918." *Army Quarterly* 8 (July 1924): 295–313.

———. "Russian Tank Doctrine." *Tank Journal* 6 (July 1939): 38–44.

———. "Tank Production." *Royal Tank Corps Journal* 5 (January 1939): 6–7.

———. "Tanks in Spain." *Royal Tank Corps Journal* 5 (January 1939): 37–39.

———. "Tanks in the Attack: French and German Tactical Theories." *Journal of the Royal United Service Institution* 83 (November 1938): 743–47.

———. "The Tank in the 'Deliberate' Attack: A Comparison Between Germany and France." *Royal Tank Corps Journal:* 5 (January 1939): 34–36.

———. "Tanks in the Next War." *Royal Tank Corps Journal* 5 (January 1939): 30–33.

———. "Tank Tactics." *Tank Journal* 6 (July 1939): 27–29.

———. "Tank Tactics in the Russian Army." *Royal Tank Corps Journal* 3 (January 1938): 37–40.

———. "Training in Egypt." *Royal Tank Corps Journal* 4 (July 1938): 54–57.

———. "Unit Training—Egypt." *Royal Tank Corps Journal* 2 (July 1937): 16–19.

———. "War: Some Recent German Ideas." *Army Quarterly* 39 (October 1939): 34–43.

———. "Western Desert Reconnaissance." *Royal Tank Corps Journal* 3 (January 1938): 24–30.

Arbutnot, Lieutenant Colonel A. G. "Horsemanship during the War." *Journal of the Royal Artillery* 46 (November 1919): 337–43.

Armstrong, Lieutenant Colonel A. G. "Army Maneuvers, 1935." *Journal of the Royal United Service Institution* 80 (November 1935): 805–12.

Atkinson, C. T. "General Liman von Sanders on His Experiences in Palestine." *Army Quarterly* 3 (January 1922): 257–75.

Baird-Smith, Lieutenant Colonel A. G. "Open War." *Army Quarterly* 5 (October 1922): 59–67.

"Bardel." "Study and Reflection v. Practical Experience." *Army Quarterly* 6 (July 1923): 318–31.

Bethell, Colonel H. A. "The Modern Attack on an Entrenched Position." *Journal of the Royal Artillery* 45 (July 1918): 121–29.

Birch, Lieutenant General Sir Noel. "Artillery Division in the Great War." *Army Quarterly* 1 (October 1920): 79–89.

Bonham-Carter, Colonel C. "The Royal Military College, Sandhurst." *Army Quarterly* 3 (October 1921): 105–13.

Brooke, Lieutenant Colonel F. "Aut Cursu, Aut Cominus Arnis." *Army Quarterly* 15 (January 1928): 259–70.

Burnett-Stuart, Major General Sir John. "The Progress of Mechanization." *Army Quarterly* 16 (April 1928): 30–51. Text originally given as a lecture at the University of London on 8 March 1928.

Charrington, Major H. V. S. "Where Cavalry Stands Today: Conclusion." *The Cavalry Journal* 17 (July 1927): 419–30.

Cherry, Lieutenant Colonel R. G. "The Aeroplane as a Long Range Gun." *Journal of the Royal Artillery* 46 (June 1919): 129–36.

Cherry, Major R. G. "The Reorganization of the Royal Regiment of Artillery with a View to Obtaining in War the Best Value from Modern Equipment with a Minimum Expenditure of Human Energy." *Journal of the Royal Artillery* 47 (November 1920): 361–78.

Clarke, F. A. S. "Ground and Mechanized Forces." *Journal of the Royal United Service Institution* 74 (August 1929): 563–66.

Clarke, Captain F. A. S. "Military Prize Essay, 1927." *Army Quarterly* 14 (July 1927): 234–53.

Clarke, Lieutenant F. A. S. "Some Further Problems of Mechanical Warfare." *Army Quarterly* 6 (July 1923): 377–85.

"Clay, Henry." "Echoes from the Mess Table and Elsewhere, VI." *Journal of the Royal Artillery* 46 (July 1919): 145–53.

———. "Echoes from the Mess Table and Elsewhere, V." *Journal of the Royal Artillery* 46 (May 1919): 49–55.

"Clericus." "The Army after the War." *Journal of the Royal United Service Institution* 63 (August 1918): 493–99.

Collins, R. J. "The Experimental Mechanized Force." *Journal of the Royal Artillery* 55 (April 1928): 12–26.

Compton, Major T. E. "The French Tanks." *Journal of the Royal United Service Institution* 65 (November 1920): 637–55.

Croft, Major W. D. "The Application of Recent Developments in Mechanics and Other Scientific Knowledge to Preparation and Training for Future War on Land." *Journal of the Royal United Service Institution* 65 (August 1920): 443–76.

Croft, Major, W. D. "Tank Minor Tactics." *Army Quarterly* 6 (April 1923): 90–98.

Cunningham, Lieutenant Colonel A. G. "The Training of the Army, 1934." *Journal of the Royal United Service Institution* 79 (November 1934): 723–32.

Dashwood-Strettell, Lieutenant Colonel C. B. "Calvalry in Open Warfare, Illustrated by the Operations Leading up to the Occupation of Mosul in November, 1918." *Journal of the Royal United Service Institution* 66 (November 1921): 598–617.

Davidson, J. H. "Passchendaele, 1917." *The Nineteenth Century and After* 89 (February 1921): 189–202.

Dening, Brevet Major B. C. "Military Prize Essay, 1924." *Army Quarterly* 8 (July 1924): 235–59.

Denning, Major B. C. "Military Prize Essay, 1929." *Army Quarterly* 18 (July 1929): 235–57.

Dewing, Major R. H. "Some Aspects of Maude's Campaign in Mesopotamia." *Army Quarterly* 14 (April 1927): 69–77.

Dimmock, Captain Lionel. "The Problem of the Tank." *Army Quarterly* 8 (July 1924): 376–80.

Ellis, Colonel Commandant Sir Hugh. "Some Notes on Tank Development during the War." *Army Quarterly* 2 (July 1921): 267–82.

Firth, Captain O. T. "Mechanized Draught and Field Artillery." *Army Quarterly* 11 (October 1925): 142–46.

Fitzpatrick, Lieutenant Colonel N. T. "An A-T Exercise." *Royal Engineers Journal* 49 (March 1935): 25–31.

Forbes, Major J. L. "The Heavy Brigade after the War." *Journal of the Royal Artillery* 47 (May 1920): 31–32.

Freybe, Captain Paul. "A German Narrative of the Battle of Cambrai." *Journal of the Royal United Service Institution* 66 (February 1921): 123–27.

Fuller, J. F. C. "The Application of Recent Developments in Mechanics and Other Scientific Knowledge to Preparation and Training for Future War on Land." *Journal of the Royal United Service Institution* 65 (May 1920): 239–74.

———. "The Development of Sea Warfare on Land and Its Influence on Future Naval Operations." *Journal of the Royal United Service Institution* 65 (May 1920): 281–98.

———. "The Foundations of the Science of War." *Army Quarterly* 1 (October 1920): 90–111.

———. "The Influence of Tanks on Cavalry Tactics." *Cavalry Journal* 10 (April 1920): 109–32; (July 1920): 307–22; (October 1920), 510–30.

———. "The Introduction of Mechanical Warfare on Land and Its Possibilities in the Near Future." *Royal Engineers Journal* 33 (January 1921): 1–13.

———. "The Mechanization of the Army." *Edinburgh Review* 248 (October 1928): 276–89.

———. "Moral, Instruction, and Leadership." *Journal of the Royal United Service Institution* 65 (November 1920): 656–74.

———. "Problems of Mechanical Warfare." *Army Quarterly* 3 (January 1922): 284–301.

———. "Science and War." *The Nineteenth Century and After* 103 (January 1928): 88–96.

———. "The Tactics of Penetration." *Journal of the Royal Artillery* 53 (July 1926): 191–221.

Germains, Victor Wallace. "'Armoured Warfare': A Plea for Common Sense." *Army Quarterly* 16 (July 1928): 363–74.

Gordon-Watson, Lieutenant D. M. L. "Has the Tank Again Made Frontal Attack Possible?" *Army Quarterly* 39 (October 1939): 154–61.

Gwinn, Major General Sir Charles W. "Army Training, 1931," *Journal of the Royal United Service Institution* 76 (November 1931): 722–30.

Harrington, Major General Sir C. H. "The Passing of an Army from War to Peace." *Journal of the Royal Artillery* 47 (May 1920): 47–66.

Hay, Major A. K. "Anti-Tank Defence." *Journal of the Royal Artillery* 47 (June 1920): 109–14.

Higgins, Lieutenant Colonel C. G. "Military Training in the Future." *Journal of the Royal United Service Institution* 65 (May 1920): 333–37.

Hilton, Captain R. "The Reorganization of the Royal Regiment of Artillery with a View to Obtaining in War the Best Value from Modern Equipment with a Minimum Expenditure of Human Energy." *Journal of the Royal Artillery* 47 (October 1920): 317–32.

Hobday, Colonel E. A. P. "Horse Management in the Field." *Journal of the Royal Artillery* 45 (May 1918): 41–59.

Holdrich, Colonel Sir T. H. "Military Education." *Journal of the Royal United Service Institution* 65 (February 1920): 116–19.

Hume, Lieutenant Colonel E. G. "Cavalry Today—'Horse v. Machine.'" *Cavalry Journal* 20 (April 1939): 180–82.

———. "Mechanical Aids to Cavalry." *Cavalry Journal* 15 (April 1925): 177–85.

Jackson, Major General Sir Louis C. "Possibilities of the Next War." *Journal of the Royal United Service Institution* 65 (February 1920): 71–89.

Johnson, Lieutenant Colonel P. "The Use of Tanks in Underdeveloped Countries." *Journal of the Royal United Service Institution* 66 (May 1921): 191–204.

Kennedy, Captain J. R. "Salisbury Plain Manoeuvres." *Army, Navy and Air Force Gazette* 75 (pt. 1, Thursday, 27 September 1934; pt. 2, Thursday, 4 October 1934): 767–68; 781–82; 789.

King, Lieutenant Colonel Norman C. "The Army after the War: Another View." *Journal of the Royal United Service Institution* 63 (November 1918): 650–56.

Knox-Nevin, H. "Should We Cut Down Our Army?" *The Nineteenth Century and After* 91 (June 1922): 1035–45.

Liddell Hart, B. H. "The Development of the 'New Model' Army: Suggestions on a Progressive, But Gradual Mechanization." *Army Quarterly* 9 (October 1924): 37–50.

———. "The 'Man-in-the-Dark' Theory of Infantry Tactics and the 'Expanding Torrent' System of Attack." *Journal of the Royal United Service Institution* 66 (February 1921): 1–22.

———. "1927—or 527?" *Journal of the Royal United Service Institution* 73 (May 1928): 268–70.

———. "The Soldier's Pillar of Fire by Night: The Need for a Framework of Tactics." *Journal of the Royal United Service Institution* 66 (November 1921): 618–26.

———. "Suggestions on the Future Development of the Combat Unit: The Tank as a Weapon of Infantry." *Journal of the Royal United Service Institution* 64 (November 1919): 666–69.

———. "The 'Ten Commandments' of the Combat Unit: Suggestions on Its Theory and Training." *Journal of the Royal United Service Institution* 64 (May 1919): 288–93.

Livingstone-Learmonth, Colonel J. E. C. "Some Notes on the Riding Establishment, Royal Horse Artillery." *Journal of the Royal Artillery* 46 (November 1919): 349–52.

Lock, Major R. F. "The Reorganization of the Royal Regiment of Artillery with a View to Obtaining in War the Best Value from Modern Equipment with a Minimum of Expenditure of Human Energy." *Journal of the Royal Artillery* 47 (February 1921): 509–19.

MacClellan, Lieutenant Colonel G. P. "Infantry Guns: Wheels or Pack?" *Journal of the Royal Artillery* 47 (November 1920): 392–94.

Macready, Colonel G. N. "The Trend of Organization in the Army." *Journal of the Royal United Service Institution* 80 (February 1935): 1–20.

Maitland-Dougall, Lieutenant Colonel. "The Future of Land Warfare." *Army Quarterly* 31 (January 1936): 262–69.

Martel, Major G. Le Q. "Small Tanks and Cavalry." *Cavalry Journal* 17 (July 1927): 437–44.

Moore, Lieutenant Colonel W. H. "The Reorganization of the Royal Regiment of Artillery with a View to Obtaining in War the Best Value from Modern Equipment with a Minimum Expenditure of Human Energy." *Journal of the Royal Artillery* 47 (August 1920): 199–212.

Morin, Lieutenant Colonel. "The Utility of Permanent Fortifications." *Journal of the Royal Artillery* 57 (January 1931): 491–505. (Translated from *Revue Militaire Française,* May 1931, by Brigadier General W. Evans.)

"Otac." "Impressions of the Armoured Force Training." *Army Quarterly* 17 (January 1929): 276–83.

Parsons, Captain H. A. J. "The Choice of Objective in War." *Army Quarterly* 13 (October 1926): 147–49.

Pemberton, Captain A. L. "The Reorganization of the Royal Regiment of Artillery with a View to Obtaining in War the Best Value from Modern Equipment with a Minimum Expenditure of Human Energy." *Journal of the Royal Artillery* 47 (September 1920): 251–64.

Phipps, Lieutenant Colonel C. F. "Heavy Brigades after the War." *Journal of the Royal Artilery* 46 (January 1920): 451–57.

Poudret, Lieutenant Colonel. "An Essay on the Employment of Cavalry." *Journal of the Royal United Service Institution* 64 (November 1919): 670–81.

"Q. L." "With the 13th Indian Cavalry Brigade in Palestine." *Journal of the Royal United Service Institution* 64 (May 1919): 232–45.

Rowan-Robinson, Lieutenant Colonel H. "Surprise, Rapidity, Concentration." *Journal of the Royal Artillery* 47 (January 1921): 473–80.

Sargeaunt, Captain W. T. "A Suggested Drill for Armoured Formations." *Journal of the Royal United Service Institution* 75 (May 1930): 375–82.

Smith, Lieutenant Colonel A. Baird. "Army Training: The Lessons of 1927." *Journal of the Royal United Service Institution* 73 (February 1921): 48–53.

Vandeleur, Lieutenant Colonel H. M. "The Reorganization of the Royal Regiment of Artillery with a View to Obtaining in War the Best Value from Modern Equipment with a Minimum of Expenditure of Human Energy." *Journal of the Royal Artillery* 47 (December 1920): 409–19.

Wade, D. A. L. "The Future of Mechanization." *Journal of the Royal United Service Institution* 74 (November 1929): 659–704.

Wake, Colonel Sir Hereward. "The Infantry Anti-Tank Weapon." *Army Quarterly* 17 (January 1929): 349–52.

———. "Mechanization and War." *Army Quarterly* 19 (January 1930): 358–64.

Wason, Major S. R. "The Reorganization of the Royal Regiment of Artillery with a View to Obtaining in War the Best Value from Modern Equipment with a Minimum of Expenditure of Human Energy." *Journal of the Royal Artillery* 47 (July 1920): 139–55.

Wavell, Lieutenant Colonel A. P. "The Strategy of the Campaign of the Egyptian Expeditionary Force." *Army Quarterly* 3 (January 1922): 235–48.

Wauchope, Brigadier General A. G. "The Destruction of the VIth Turkish Army." *Journal of the Royal United Service Institution* 64 (August 1919): 436–59.

Weber, Lieutenant Colonel W. H. F. "The Development of Mobile Artillery, 1914–1918." *Journal of the Royal United Service Institution* 64 (February 1919): 49–58.

White, Major O. W. "The Abolition of Horse Transport in the Administrative Services." *Journal of the Royal United Service Institution* 66 (February 1921): 58–80.

Published Sources: Secondary

Books

Barnett, Correlli. *Britain and Her Army.* London: Allan Lane The Penguin Press, 1970.

———. *The Collapse of British Power.* New York: William Morrow & Co., Inc., 1972.

———. *The Desert Generals.* New York: Viking Press, 1961.

Bond, Brian. *British Military Policy between the Two World Wars.* Oxford: Clarendon Press, 1980.

———. *The Victorian Army and the Staff College.* London: Eyre McThuen, 1972.

———. *Liddell Hart: A Study of His Military Thought.* New Brunswick, N.J.: Rutgers University Press, 1977.

Butler, Sir J[ames] R. M. *Grand Strategy,* vol. 2, *September 1939–June 1941,* in *History of the Second World War, United Kingdom Military Series,* Sir J[ames] R. M. Butler, gen. ed. London: HMSO, 1957.

Carver, Field Marshal Lord Michael. *The Apostles of Mobility: The Theory and Practice of Armoured Warfare.* London: Holmes and Meier, 1979.

Chamberlain, Peter, and Chris Ellis. *British and American Tanks of World War II.* New York: Arco Publishing Company, Inc., 1969.

Cole, Howard N. *The Story of Aldershot.* Aldershot: Gale and Polden, Ltd., 1951.

Dennis, Peter. *Decision by Default: Peacetime Conscription and British Defence, 1919–1939.* Durham, N.C.: Duke University Press, 1972.

Edmonds, Brigadier General Sir James E. *Military Operations France and Belgium: 1916, to July 1 the Somme* in *History of the Great War Based on Official Documents,* Brigadier General Sir James E. Edmonds, gen. ed. London: Macmillan and Co., Ltd., 1932.

———. *Military Operations France and Belgium: 1918, August 8–Septetmber 26* in *History of the Great War Based on Official Documents,* Brigadier General Sir James E. Edmonds, gen. ed. London: Macmillan and Co., Ltd., 1947.

Ellis, Major [Lionel] F. *The War in France and Flanders, 1939–40,* in *History of the Second World War, United Kingdom Military Serieds,* Sir J[ames] R. M. Butler, gen. ed. London: HMSO, 1953.

Falls, Captain Cyril. *Military Operations France and Belgium: 1917, The German Retreat to the Hindenburg Line and the Arras Battles,* in *History of the Great War Based on Official Documents,* Brigadier General Sir James E. Edmonds, gen. ed. London: Macmillan and Co., Ltd., 1940.

Fussell, Paul. *The Great War and Modern Memory.* New York & London: Oxford University Press, 1975.

Godwin-Austen, Major A. R. *The Staff and the Staff College.* London: Constable & Co., Ltd., 1927.

Graves, Robert, & Alan Hodge. *The Long Week-End: A Social History of Great Britain, 1918–1939.* New York: W. W. Norton & Co., 1963.

Guggisberg, Captain F. G. *"The Shop": The Story of the Royal Military Academy.* 2nd ed. London: Cassell & Co., Ltd., 1902.

Hancock, Sir W. Keith, and M. M. Gowing. *British War Economy,* in *History of the Second World War, United Kingdom Civil Series,* Sir W. Keith Hancock, gen. ed. London: HMSO, 1949.

Higham, Robin. *Armed Forces in Peacetime: Britain, 1918–1940.* Hamden, Conn.: Anchor Books, 1962.

———. *The Military Intellectuals in Britain: 1918–1939.* New Brunswick, N.J.: Rutgers University Press, 1966.

Horne, Alistaire. *To Lose a Battle: France, 1940.* London: Macmillan & Co., Ltd., 1969.

Jackson, W. F. G. *The Battle for North Africa, 1940–1943.* New York: Mason/Charter, 1975.

Lewin, Ronald. *Man of Armour: A Study of Lieutenant General Vyvyan Pope.* London: Leo Cooper, 1976.

Luvaas, Jay. *The Education of an Army.* London: Cassell, 1965.

Macksey, Kenneth. *Armoured Crusader.* London: Hutchinson, 1967.

Marshall, S. L. A. *Blitzkrieg.* New York: William Morrow & Co., 1940.

Miles, Captain Wilfred. *Military Operations France and Belgium: 1916, July 2 to the End of the Somme Battles.* in *History of the Great War Based on Official Documents,* Brigadier General Sir James E. Edmonds, gen. ed. London: Macmillan and Co., Ltd., 1938.

Mockler-Ferryman, Major A. F. *Annals of Sandhurst.* London: William Heinemann, 1900.

Mowat, Charles Lock. *Britain Between the Wars, 1918–1940.* Chicago: University of Chicago Press, 1963.

Ogorkiewicz, Richard N. *Armoured Forces.* London: Arms and Armour Press, 1970.

Playfair, Major General I[an] S.O. *The Mediterranean and Middle East,* vol 2: *The Germans Come to the Help of their Ally,* in *History of the Second World War, United Kingdom Military Series,* Sir J[ames] R. M. Butler, gen. ed. London: HMSO, 1956.

———. *The Mediterranean and Middle East,* vol 3: *British Fortunes Reach Their Lowest Ebb,* in *History of the Second World War, United Kingdom Military Series,* Sir J[ames] R. M. Butler, gen. ed. London: HMSO, 1960.

Postan, M[ichael] M. *British War Production,* in *History of the Second World War, United Kingdom Civil Series,* Sir Keith Hancock, gen. ed. London: HMSO, 1952.

Postan, M[ichael] M., D. Hay, and J. D. Scott. *The Design and Development of Weapons,* in *History of the Second World War, United Kingdom Civil Series,* Sir Keith Hancock, gen. ed. London: HMSO, 1964.

Preston, Richard A., Sydney F. Wise, and Herman O. Werner. *Men in Arms.* New York: Frederick A. Praeger, 1962.

Ropp, Theodore. *War in the Modern World.* Durham, N.C.: Duke University Press, 1959.

Shay, Robert Paul. *British Rearmament in the Thirties.* Princeton, N.J.: Princeton University Press, 1977.

Sheppard, Eric Williams. *A Short History of the British Army,* 4th ed. London: Constable and Co., Ltd., 1950.

Smyth, Brigadier Sir John. *Sandhurst.* London: Weindenfeld & Nicolson, 1961.

Taylor, A. J. P. *English History, 1914–1945.* New York: Oxford University Press, 1965.

Terraine, John. *Ordeal of Victory.* Philadelphia & New York: J. B. Lippincott Co., 1963.

———. *The Western Front, 1914–1918.* London: Hutchinson & Co., Ltd., 1964.

Thomas, Hugh. *The Story of Sandhurst.* London: Hutchinson, 1961.

Trythall, Anthony. *"Boney" Fuller.* New Brunswick, N.J.: Rutgers University Press, 1977.

Turner, Ernest S. *Gallant Gentlemen: A Portrait of the British Officer, 1660–1956.* London: Michael Joseph, 1956.

Vagts, Alfred. *A History of Militarism.* New York: Free Press, 1967.

Watt, Donald C. *Too Serious a Business.* Berkeley, Calif.: University of California Press, 1975.

Wheldon, John. *Machine Age Armies.* London: Adelard-Schuman, 1969.

Whitehouse, Arch. *Tank.* Garden City, N.Y.: Doubleday & Co., Inc., 1960.

Articles

Barnett, Correlli. "The Education of Military Elites." *Journal of Contemporary History* 2 (July 1967): 15–36.

Bond, Brian. "The First World War." In *The Shifting Balance of World Forces, 1898–1945,* edited by C. L. Mowat. The New Cambridge Modern History. Vol. 12. Cambridge: The University Press, 1968.

Gibbs, Norman. "British Strategic Doctrine, 1918–1939." In *The Theory and Practice of War,* edited by Michael Howard. London: Cassell, 1965.

Gibson, Irvin M. "Maginot and Liddell Hart: The Doctrine of Defence." In *The Makers of Modern Strategy: Military Thought from Machiavelli to Hitler,* edited Edward M. Earle. New York: Antheneum, 1967.

Howard, Michael. "The Armed Forces." In *Material Progress and World-Wide Problems, 1870–98,* edited by F. H. Hinsley. The New Cambridge Modern History. Vol. 11. Cambridge: Cambridge University Press, 1967.

———. "Strategy and Policy in Twentieth Century Warfare." In his *Studies in War and Peace.* London: Temple Smith, 1970.

Luvaas, Jay. "European Military Thought and Doctrine, 1870–1914." In *The Theory and Practice of War,* edited by Michael Howard. New York: Frederick A. Praeger, 1966.

Maude, P. D. "Lord Haldane's Reorganization of the British Army, 1905–1912." *The Army Quarterly* 55 (October 1947): 47–57.

Paret, Peter. "Clausewitz and the Nineteenth Century." In *The Theory and Practice of War,* edited by Michael Howard. London: Cassell & Co., Ltd., 1965.

Pile, General Sir Frederick. "Liddell Hart and the British Army." In *The Theory and Practice of War,* edited by Michael Howard. New York: Frederick A. Praeger, 1966.

Poe, Brice, II. "British Army Reforms, 1902–1914." *Military Affairs* 31 (Fall 1967): 131–38.

Possony, Stefan T., and Etienne Mantoux. "DuPicq and Foch: The French School." *The Makers of Modern Strategy: Miltiary Thought from Machiavelli to Hitler,* edited by Edward M. Earle. New York: Antheneum, 1967.

Razzell, P. E. "Social Origins of Officers in the Indian and British Home Army." *British Journal of Sociology* 14, no. 3 (September 1962): 248–60.

Rothfels, Hans. "Clausewitz." In *The Makers of Modern Strategy,* edited by Edward M. Earle. New York: Antheneum, 1967.

Index